T0269289

SpringerBriefs in Mathematics

SpringerBriefs in Mathematics showcases expositions in all areas of mathematics and applied mathematics. Manuscripts presenting new results or a single new result in a classical field, new field, or an emerging topic, applications, or bridges between new results and already published works, are encouraged. The series is intended for mathematicians and applied mathematicians.

BCAM SpringerBriefs

BCAM *SpringerBriefs* aims to publish contributions in the following disciplines: Applied Mathematics, Finance, Statistics and Computer Science. BCAM has appointed an Editorial Board, who evaluate and review proposals.

Typical topics include: a timely report of state-of-the-art analytical techniques, bridge between new research results published in journal articles and a contextual literature review, a snapshot of a hot or emerging topic, a presentation of core concepts that students must understand in order to make independent contributions.

Please submit your proposal to the Editorial Board or to Francesca Bonadei, Executive Editor Mathematics, Statistics, and Engineering: francesca.bonadei@springer.com

More information about this series at http://www.springer.com/series/10030

Diego Ricciotti

p-Laplace Equation in the Heisenberg Group

Regularity of Solutions

Diego Ricciotti
Department of Mathematics
University of Pittsburgh
Pittsburgh, PA
USA

ISSN 2191-8198   ISSN 2191-8201   (electronic)
SpringerBriefs in Mathematics
ISBN 978-3-319-23789-3   ISBN 978-3-319-23790-9   (eBook)
DOI 10.1007/978-3-319-23790-9

Library of Congress Control Number: 2015958335

Printed on acid-free paper

This Springer imprint is published by SpringerNature
The registered company is Springer International Publishing AG Switzerland

We are the people
we rule the world

Preface

This work is based on my master's thesis from the University of Bologna, written under the supervision of my advisors Juan Manfredi and Bruno Franchi, and is intended to present a self-contained introduction to the p-Laplace equation and related regularity theory in the Heisenberg group. We also obtain new regularity results in the nondegenerate case $1 < p < 2$. I thank Drs. Franchi and Manfredi for encouraging me to prepare this manuscript and to submit it to the BCAM Springer Briefs.

The regularity theory of nonlinear elliptic equations is quite well understood in the Euclidean case, but in many problems in science and engineering the most natural setup is subelliptic. This is the case in non-holonomic mechanics, nonlinear elasticity, robotic control theory, and certain models in the neurobiology of vision, to name but a few examples. In the subelliptic case the velocity fields are restricted to reflect non-holonomic constraints, and this leads to the study of non-commuting vector fields generating nilpotent Lie algebras. The simplest, yet very important, example is the Heisenberg group, which we describe in Chap. 2. The complications resulting from the lack of commutativity of the primary vector fields generate new challenges in regularity theory, especially in the nonlinear case but also in the linear case, where the key result was proved by Hormander.

When we minimize non-quadratic energy functionals, the resulting Euler equations are quasilinear, of p-Laplacian type, as described in Chap. 3. The expected regularity would be that solutions have Holder continuous derivatives. In Chap. 5 we present Zhong's Lipschitz continuity results for p-harmonic functions in the full range $1 < p < \infty$. In the nondegenerate case, where vanishing gradients do not present a difficulty, we would expect C^∞ regularity. This was known for the case $p \geq 2$. Our main new result presented in this manuscript is a proof of this fact valid for the full range $1 < p < \infty$.

Pittsburgh   Diego Ricciotti
May 2015

Acknowledgments

I would like to thank my thesis advisors Bruno Franchi, for all his support during my years in Bologna, and Juan Manfredi, for all the time he has dedicated to me since before my arrival in Pittsburgh.

I wish to express my appreciation to Prof. Andras Domokos, whose answers and comments have been a great help in the development of this work.

I am also particularly grateful to Prof. Giovanna Citti, coordinator of the CAP EXCEL exchange program, for her help and guidance.

Contents

Acronyms

∇	Euclidean gradient
∇^2	Euclidean Hessian
$\| \cdot \|_E$	Euclidean norm
$\langle \, \cdot \, \rangle$	Euclidean scalar product
$B_r^E(x)$	Euclidean ball of radius r and center x
$\mathbb{H}$	Heisenberg group
$\mathfrak{g}$	Lie algebra associated to $\mathbb{H}$
$C_0^\infty(\Omega)$	Smooth functions with compact support in Ω
$\nabla_{\mathbb{H}}$	Horizontal gradient
$\mathrm{div}_{\mathbb{H}}$	Horizontal divergence
$\nabla_{\mathbb{H}}^2$	Horizontal Hessian
$HW^{1,p}(\Omega)$	Horizontal Sobolev space
$HW_{loc}^{1,p}(\Omega)$	Local horizontal Sobolev space
$HW_0^{1,p}(\Omega)$	Closure of $C_0^\infty(\Omega)$ in the horizontal Sobolev norm
d_{cc}, d	Carnot–Carathéodory distance
$B_r(x)$	Carnot–Carathéodory ball of radius r and center x
$\Gamma^\alpha(\Omega)$	Hölder functions with respect to d_{cc}
$\| \cdot \|_K$	Korányi norm
supp	Support of a function
$\Omega_r(x)$	$= \Omega \cap B_r(x)$
$M^{p,\lambda}(\Omega)$	Morrey space
$\mathcal{L}^{p,\lambda}(\Omega)$	Campanato space
$\mathcal{D}_p(u)$	p-Dirichlet functional
$\fint_\Omega$	Integral average over Ω
$\mathrm{Lie}(X)$	Lie algebra generated by the set of vector fields X

$\Delta_{Z,h}^{\alpha}$ First-order difference quotient of exponent α in the direction of the vector field Z

$\Delta_{Z,h}$ First-order difference quotient of exponent 1 in the direction of the vector field Z

$\Delta_{Z,h}^{2,\alpha}$ Second-order difference quotient of exponent α in the direction of the vector field Z

Chapter 1
Introduction

Abstract In this introductory chapter we present and motivate the result of this work concerning regularity of solutions to the p-Laplace equation in the Heisenberg group and we give an overview of some previous directly related results.

Keywords p-Laplace equation · Heisenberg group · Regularity

The Heisenberg group $\mathbb{H}$ is the simplest example of Carnot group, i.e. a connected, simply connected nilpotent Lie group $\mathbb{G}$ whose associated Lie algebra $\mathfrak{g}$ admits a finite stratification. To be more precise we can identify, via the Exponential map, $\mathbb{G}$ with the Lie group $(\mathbb{R}^n, *)$ where $*$ is in general a non commutative group operation. The Lie algebra $\mathfrak{g}$ admits a stratification in that it can be written as a direct sum of linear subspaces $\mathfrak{g} = \bigoplus_{i=1}^{k} V_i$ such that $[V_1, V_i] = V_{i+1}$ where $V_{k+1} = 0$.

For the first Heisenberg group $\mathbb{H} = (\mathbb{R}^3, *)$ indicating points $x, y \in \mathbb{H}$ by $x = (x_1, x_2, z)$ and $y = (y_1, y_2, s)$ the group operation is

$$x * y = (x_1, x_2, z) * (y_1, y_2, s) = \left(x_1 + y_1, x_2 + y_2, z + s + \frac{1}{2}(x_1 y_2 - x_2 y_1) \right)$$

(1.1)

and a basis of left-invariant vector fields for the associated Lie algebra $\mathfrak{h}$ is given by

$$X_1 = \partial_{x_1} - \frac{x_2}{2}\partial_z, \quad X_2 = \partial_{x_2} + \frac{x_1}{2}\partial_z \quad \text{and} \quad T = \partial_z.$$

(1.2)

If $u : \Omega \longrightarrow \mathbb{R}$ is a function from an open subset of $\mathbb{H}$ we indicate by $\nabla_{\mathbb{H}} u = (X_1 u, X_2 u)$ the horizontal gradient of u.

This work deals with regularity o f weak solutions to the p-Laplace equation in the first Heisenberg group

$$\sum_{i=1}^{2} X_i \left(\left(\delta^2 + |\nabla_{\mathbb{H}} u|^2 \right)^{\frac{p-2}{2}} X_i u \right) = 0 \quad \text{in } \Omega$$

(1.3)

which is the subelliptic counterpart of the Euclidean p-Laplace equation

© The Author(s) 2015
D. Ricciotti, *p-Laplace Equation in the Heisenberg Group*,
SpringerBriefs in Mathematics, DOI 10.1007/978-3-319-23790-9_1

$$\sum_{i=1}^{2} \partial_i \left(\left(\delta^2 + |\nabla u|^2 \right)^{\frac{p-2}{2}} \partial_i u \right) = 0 \quad \text{in } \Omega. \tag{1.4}$$

Equations (1.3) and (1.4) are respectively the Euler-Lagrange equations for the p-Dirichlet functionals

$$\mathcal{D}_p(u) = \frac{1}{p} \int_{\Omega} \left(\delta^2 + |\nabla_{\mathbb{H}} u|^2 \right)^{\frac{p}{2}} \, dx \tag{1.5}$$

and

$$\mathcal{D}_p(u) = \frac{1}{p} \int_{\Omega} \left(\delta^2 + |\nabla u|^2 \right)^{\frac{p}{2}} \, dx. \tag{1.6}$$

It is customary to set this problem in the Sobolev space $W^{1,p}(\Omega)$ where it is easy to prove existence and uniqueness results and then try to recover the regularity of the solution. The same thing can be done in the Heisenberg group setting, where we have horizontal Sobolev spaces $HW^{1,p}(\Omega)$ of all L^p functions whose horizontal derivatives are in L^p. We say that $u \in HW^{1,p}(\Omega)$ is a weak solution of Eq. (1.3) if

$$\int_{\Omega} \sum_{i=1}^{2} \left(\delta^2 + |\nabla_{\mathbb{H}} u|^2 \right)^{\frac{p-2}{2}} X_i u X_i \varphi dx = 0 \quad \text{for all } \varphi \in C_0^{\infty}(\Omega). \tag{1.7}$$

The theory in the Euclidean case is well developed. The p-Laplace equation is a generalization of the classical Laplace equation which is the model for all elliptic linear equations. For $p = 2$ the p-Laplacian coincides with the usual Laplace operator, but for $p \neq 2$ the p-Laplace operator is non linear and degenerate if $p > 2$ or singular if $1 < p < 2$ where the gradient vanishes. The p-Laplace equation is not only relevant in Mathematical Analysis but also in the theory of quasi-conformal maps [4] (when p is equal to the dimension of the space in the Euclidean case or the homogeneous dimension in the case of Carnot groups), in game theory and also in numerous fields of Physics, such as non-Newtonian fluid mechanics [2], fluid flows through porous media [1, 24], non-linear elasticity [21] to quote only a few.

Concerning regularity matters, Ural'tseva [14] in 1968 proved that weak solutions are in $C_{loc}^{1,\alpha}(\Omega)$ for $p \geq 2$, and Lewis [15] and DiBenedetto [6] proved independently in 1983 the case $1 < p < 2$ with two different approaches. This result is in general optimal as some counterexamples can be found. For the case of systems of differential equations we quote the work of Uhlenbeck [23]. In the plane the optimal regularity $C_{loc}^{k,\alpha}(\Omega) \cap W_{loc}^{k+2,2}(\Omega)$ was established by Iwaniec and Manfredi [12] where they give an explicit formula relating k and α.

Typically, in order to prove regularity results, we need to differentiate the original equation to prove that the derivatives of the solution satisfy other particular linear equations. In the Heisenberg group the main difficulty is that when we try to differentiate the equation some extra terms containing the vertical derivative Tu appear, and we don't have a priori estimates because we only assume $u \in HW^{1,p}(\Omega)$. Therefore we need to control the vertical terms first.

The theory in the Heisenberg groups starts with a paper that goes back to Hörmander [11] where the linear case is considered.

The Hölder regularity of solutions of equations modeled on (1.3) was established by Capogna and Garofalo [5] and Lu [16].

Capogna studies in [3] the $C^{1,\alpha}$ regularity for subelliptic quasi-linear equations

$$\sum_{i=1}^{n} X_i a_i(x, \nabla_{\mathbb{H}} u) = f(x) \tag{1.8}$$

with

$$\sum_{i,j=1}^{2} \partial_{z_j} a_i(x, z)\xi_i \xi_j \geq c|\xi|^2 \tag{1.9}$$

$$|\partial_{z_j} a_i(x, z)| \leq C \tag{1.10}$$

$$|\nabla_x a_i(x, z)| \leq C(1 + |z|). \tag{1.11}$$

This allows him to prove C^∞ regularity for the solutions of the p-Laplace equation for $p \geq 2$ under the additional assumption

$$0 < M^{-1} < |\nabla_{\mathbb{H}} u| < M,$$

i.e. for a non degenerate case. He extends to the non linear setting a technique used by Kohn [13] and he uses differential quotients defined in terms of the Fourier transform together with a result by Peetre [22] to get $Tu \in L^2$ as a first step.

For the case $p \neq 2$ some regularity results have been established by Manfredi and Domokos [8, 9] via the Cordes perturbation technique for p near 2 without explicit bounds on p, but these are valid also in the degenerate case. Later Manfredi and Mingione [17] were able to prove $C^{1,\alpha}$ regularity in the non degenerate case for $2 \leq p < c(n) < 4$ and by adapting an argument used by Capogna prove C^∞ regularity for this range of values of p. An essential starting point is $Tu \in L^p$ proved by Domokos for $1 < p < 4$ in [7] where he uses difference quotients techniques to gain regularity for both vertical and horizontal derivatives, extending results by Marchi [18, 19].

In [20] Mingione, Zatorska-Goldstein and Zhong improve the bound on p overcoming the dependence on the dimension of the space and establish $C^{1,\alpha}$ regularity for $2 \leq p < 4$. They also prove Lipschitz continuity for the solutions of the degenerate equation (always in the range $2 \leq p < 4$). However the restriction on the values of p seems somehow unnatural, as some results along this line of thought are available (see for instance Garofalo [10] in which he shows $C^{1,\alpha}$ regularity for solutions with some special symmetries for $p \geq 2$).

In this monograph we present a self contained proof of the Lipschitz regularity of solutions to the degenerate p-Laplace equation (1.3) following the approach of Domokos [7], which avoids explicit use of the Fourier transform:

Theorem 1.1 *Let $u \in HW^{1,p}(\Omega)$, $1 < p < \infty$ be a weak solution of the degenerate p-Laplace equation (1.3). Then*

$$\|\nabla_{\mathbb{H}} u\|_{L^{\infty}(B_r)} \le C_p \left(\fint_{B_{2r}} |\nabla_{\mathbb{H}} u|^p \mathrm{d}x \right)^{\frac{1}{p}} \tag{1.12}$$

for every ball B_r such that the concentric ball $B_{2r} \subset \Omega$.

This theorem was proved by Zhong (see [25], Theorem 1.1) by extending the Hilbert-Haar existence theory to the Heisenberg group and using Capogna's results in [3] to get the proof started.

Solutions to the non degenerate p-Laplace equation (1.3) are C^{∞} smooth as proved by Capogna in [3] for $p \ge 2$. Building on techniques developed by Domokos in [7] in this monograph we prove this theorem for the full range $1 < p < \infty$:

Theorem 1.2 *Let $u \in HW^{1,p}(\Omega)$, $1 < p < \infty$ be a weak solution of the non-degenerate p-Laplace equation (1.3). Then $u \in C^{\infty}(\Omega)$.*

In Chap. 2 we describe some features of Carnot groups, with a particular focus on the Heisenberg group, and present some useful properties that we will need. In particular subelliptic versions of Poincaré and Sobolev inequality and results on fractional difference quotients will play a fundamental role in the proofs of the following sections. In fact, there is a characterization of Sobolev spaces in terms of difference quotients and the cited inequalities will be necessary for the Moser's iteration needed to prove the desired results.

In Chap. 3 we describe in detail some features of the p-Laplace equation and relations with variational problems, establishing existence and uniqueness results for the relative Dirichlet problem and comparison principle. At the end we present the so called Hilbert-Haar existence theory, which demonstrates that solutions of Eq. (1.3) are Lipschitz continuous in certain types of domains satisfying a particular convexity condition. This is a focal point because it simplifies the calculations and allows us to use Capogna's results [3] without having to assume a priori the boundedness of the horizontal gradient of solutions.

Chapter 4 is where we give a proof of Theorem 1.2. Once we have that $\|\nabla_{\mathbb{H}} u\|_{L^{\infty}} \le M$ we can adapt Capogna's [3] proofs, but we avoid the use of Fourier transform and instead follow Domokos [7]. In particular we use fractional difference quotients of second order to prove summability results for the first order difference quotients with increasing fractional exponent until we are able to prove summability of the vertical derivative $Tu \in HW^{1,2}$. This will lead to $\nabla_{\mathbb{H}} u \in HW^{1,2}$ and consequently we will be able to differentiate the equation and obtain that Tu and $\nabla_{\mathbb{H}} u$ satisfy certain linear subelliptic equations, whose theory is well known in the literature and gives us $C^{1,\alpha}$ regularity. By classical bootstrap arguments using subelliptic version of Schauder estimates we can achieve smoothness of solutions. However the bounds obtained depend on the non degeneracy parameter δ preventing the passage to the limit to get regularity for the degenerate case.

In Chap. 5 we will prove Theorem 1.1. In order to do this we will establish several Caccioppoli type estimates with constants independent of the non degeneracy parameter δ. The main result here is Lemma 5.3 that allows us to absorb the vertical derivative that we cannot control a priori. We will then run a suitable version of Moser iteration and eventually passing to the limit we will conclude Lipschitz regularity for the degenerate case.

References

1. Aronson, D.G.: The porous medium equation. Nonlinear Diffusion Problems (Montecatini Terme, 1985), Volume 1224 of Lecture Notes in Mathematics, pp. 1–46. Springer, Berlin (1986)
2. Astrita, G., Marrucci, G.: Principles of Non-Newtonian Fluid Mechanics. McGraw-Hill, New York (1974)
3. Capogna, L.: Regularity of quasi-linear equations in the Heisenberg group. Comm. Pure Appl. Math. **50**(9), 867–889 (1997)
4. Capogna, L.: Regularity for quasilinear equations and 1-quasiconformal maps in Carnot groups. Math. Ann. **313**(2), 263–295 (1999)
5. Capogna, L., Garofalo, N.: Regularity of minimizers of the calculus of variations in Carnot groups via hypoellipticity of systems of Hörmander type. J. Eur. Math. Soc. (JEMS) **5**(1), 1–40 (2003)
6. DiBenedetto, E.: $C^{1+\alpha}$ local regularity of weak solutions of degenerate elliptic equations. Nonlinear Anal. **7**(8), 827–850 (1983)
7. Domokos, A.: Differentiability of solutions for the non-degenerate p-Laplacian in the Heisenberg group. J. Differ. Equ. **204**(2), 439–470 (2004)
8. Domokos, A., Manfredi, J.J.: $C^{1,\alpha}$-Regularity for p-harmonic Functions in the Heisenberg Group for p Near. The p-harmonic Equation and Recent Advances in Analysis, Volume 370 of Contemporary Mathematics, pp. 17–23. American Mathematical Society, Providence (2005)
9. Domokos, A., Manfredi, J.J.: Subelliptic Cordes estimates. Proc. Am. Math. Soc. **133**(4), 1047–1056 (2005). (electronic)
10. Garofalo, N.: Gradient bounds for the horizontal p-Laplacian on a Carnot group and some applications. Manuscripta Math. **130**(3), 375–385 (2009)
11. Hörmander, L.: Hypoelliptic second order differential equations. Acta Math. **119**, 147–171 (1967)
12. Iwaniec, T., Manfredi, J.J.: Regularity of p-harmonic functions on the plane. Rev. Mat. Iberoam. **5**(1–2), 1–19 (1989)
13. Kohn, J.J.: Pseudo-differential operators and hypoellipticity. In: *Partial differential equations (Proc. Sympos. Pure Math., Vol. XXIII, Univ. California, Berkeley, Calif., 1971)*, pp. 61–69. American Mathematical Society, Providence (1973)
14. Ladyzhenskaya, O.A., Ural'tseva, N.N.: Linear and Quasilinear Elliptic Equations. Translated from the Russian by Scripta Technica, Inc. Translation editor: Leon Ehrenpreis. Academic Press, New York (1968)
15. Lewis, J.L.: Regularity of the derivatives of solutions to certain degenerate elliptic equations. Indiana Univ. Math. J. **32**(6), 849–858 (1983)
16. Lu, G.: Embedding theorems into Lipschitz and BMO spaces and applications to quasilinear subelliptic differential equations. Publ. Mat. **40**(2), 301–329 (1996)

17. Manfredi, J.J., Mingione, G.: Regularity results for quasilinear elliptic equations in the Heisenberg group. Math. Ann. **339**(3), 485–544 (2007)
18. Marchi, S.: $C^{1,\alpha}$ local regularity for the solutions of the p-Laplacian on the Heisenberg group for $2 \leq p < 1 + \sqrt{5}$. Z. Anal. Anwendungen **20**(3), 617–636 (2001)
19. Marchi, S.: $C^{1,\alpha}$ local regularity for the solutions of the p-Laplacian on the Heisenberg group. The case $1 + \frac{1}{\sqrt{5}} < p \leq 2$. Comment. Math. Univ. Carolin. **44**(1), 33–56 (2003)
20. Mingione, G., Zatorska-Goldstein, A., Zhong, X.: Gradient regularity for elliptic equations in the Heisenberg group. Adv. Math. **222**(1), 62–129 (2009)
21. Ôtani, M.: A remark on certain nonlinear elliptic equations. Proc. Fac. Sci. Tokai Univ. **19**, 23–28 (1984)
22. Peetre, J.: A theory of interpolation of normed spaces. Notas de Matemática, No. 39. Instituto de Matemática Pura e Aplicada, Conselho Nacional de Pesquisas, Rio de Janeiro (1968)
23. Uhlenbeck, K.: Regularity for a class of non-linear elliptic systems. Acta Math. **138**(3–4), 219–240 (1977)
24. Vázquez, J.L.: The Porous Medium Equation. Oxford Mathematical Monographs. The Clarendon Press Oxford University Press, Oxford (2007) (Mathematical theory)
25. Zhong, X.: Regularity for variational problems in the Heisenberg group. preprint (2009)

Chapter 2
The Heisenberg Group

Abstract This chapter is meant to give a brief and by no means complete description of the Heisenberg group $\mathbb{H}$, that will be the setting of this work. Customarily this group is presented as a particular group on $\mathbb{R}^3$. This is not restrictive and to explain why we recall some definitions and basic properties of Carnot groups in order to make the exposition self-contained. We refer to the monograph (Bonfiglioli et al., Stratified Lie Groups and Potential Theory for their Sub-Laplacians, 2007 [1]) for a complete presentation of Carnot groups.

Keywords Carnot groups · Heisenberg group · Difference quotients

2.1 Carnot Groups

Definition 2.1 A Carnot Group is a connected, simply connected Lie group $(\mathbb{G}, *)$ whose associated Lie algebra $\mathfrak{g}$ admits a stratification, namely a decomposition as direct sum of linear subspaces V_i, $i = 1, \ldots, r$ such that $[V_1, V_i] = V_{i+1}$ with $V_{r+1} = 0$.

Remark 2.1 The stratification implies that the Lie algebra $\mathfrak{g}$ is nilpotent of step r. We also say that the Lie group $\mathbb{G}$ is nilpotent.

Remark 2.2 The step r and the dimensions n_i of the linear spaces V_i are independent from the stratification. Whenever we will talk about a Carnot group $\mathbb{G}$ we will consider a fixed stratification $\mathfrak{g} = \bigoplus_{i=1}^{r} V_i$.

Definition 2.2 A basis $B = (X_1^1, X_2^1, \ldots, X_{n_1}^1, X_1^2, \ldots, X_{n_2}^2, \ldots, X_1^r, \ldots, X_{n_r}^r)$ of $\mathfrak{g}$ is called adapted to the stratification if $(X_1^i, \ldots, X_{n_i}^i)$ is a basis of V_i for $i = 1, \ldots, r$.

Definition 2.3 We call homogeneous dimension of $\mathbb{G}$ the number

$$Q = \sum_{i=1}^{r} i\, n_i. \tag{2.1}$$

© The Author(s) 2015
D. Ricciotti, *p-Laplace Equation in the Heisenberg Group*,
SpringerBriefs in Mathematics, DOI 10.1007/978-3-319-23790-9_2

We denote by $e^{tX}x = \Gamma_X(t, x)$ the integral curve of the vector field $X \in \mathfrak{g}$ at time t starting at the point $x \in \mathbb{G}$. There is a canonical way to associate elements of the Lie algebra $\mathfrak{g}$ to elements of the group $\mathbb{G}$ using integral curves.

Definition 2.4 (*Exponential map*) The exponential map is

$$\exp :\mathfrak{g} \longrightarrow \mathbb{G}$$
$$\exp(X) = \Gamma_X(1, e)$$

where we have denoted by e the identity element of the group $\mathbb{G}$.

In general $\exp$ is a diffeomorphism of a neighbourhood of $0 \in \mathfrak{g}$ into a neighbourhood of $e \in \mathbb{G}$ but in Carnot groups we have a stronger result.

Theorem 2.1 *If $\mathbb{G}$ is a Carnot group then $\exp$ is a global diffeomorphism between $\mathfrak{g}$ and $\mathbb{G}$.*

We will now briefly describe how we can identify a Carnot group $\mathbb{G}$ with a Carnot group on $\mathbb{R}^n$ satisfying certain properties. First we recall the celebrated Baker–Campbell–Hausdorff formula.

Theorem 2.2 (Baker–Campbell–Hausdorff formula) *Let $(\mathbb{G}, *)$ be a Carnot group with Lie algebra $\mathfrak{g}$. Then*

$$\exp(X) * \exp(Y) = \exp(X \diamond Y) \tag{2.2}$$

for every $X, Y \in \mathfrak{g}$. We have denoted by $\diamond$ the following operation:

$$X \diamond Y = \sum_{n=1}^{\infty} \frac{(-1)^{n+1}}{n} \sum_{\substack{p_i+q_i \geq 1 \\ 1 \leq i \leq n}} \frac{(\mathrm{ad}X)^{p_1}(\mathrm{ad}Y)^{q_1} \cdots (\mathrm{ad}X)^{p_n}(\mathrm{ad}Y)^{q_n-1}Y}{(\sum_{j=1}^{n}(p_j + q_j))p_1!q_1! \ldots p_n!q_n!} \tag{2.3}$$

and $\mathrm{ad}X : \mathfrak{g} \longrightarrow \mathfrak{g}$, $(\mathrm{ad}X)Y = [X, Y]$ is the adjoint operator.

Remark 2.3 Since the algebra $\mathfrak{g}$ is nilpotent the previous series is indeed a finite sum. Theorem (2.2) holds in a more general setting with some modification. Here we only state results adapted to case of Carnot groups wich are the ones we are interested in. We refer to [5] for more exhaustive results in general Lie groups.

Theorem 2.3 *Let $(\mathbb{G}, *)$ be a Carnot group with Lie algebra $\mathfrak{g}$. Then $(\mathfrak{g}, \diamond)$ is a Lie group isomorphic to $(\mathbb{G}, *)$ via the exponential map.*

Now if we choose a basis for the vector space $\mathfrak{g}$ we can identify the Lie group $(\mathfrak{g}, \diamond)$ with the Lie group $(\mathbb{R}^n, \star)$ where n is the dimension of $\mathfrak{g}$ and $\star$ is a suitable group operation. More precisely we have the following result:

Theorem 2.4 *Let $(\mathbb{G}, *)$ be a Carnot group. Then there exist a Lie group $(\mathbb{R}^n, \star)$ which is isomorphic to $(\mathbb{G}, *)$.*

Proof From Theorem 2.3 we know that the Lie algebra $\mathfrak{g}$ can be endowed with the operation $\diamond$ defined in (2.3) making $(\mathfrak{g}, \diamond)$ a Lie group and

$$\exp : (\mathfrak{g}, \diamond) \longrightarrow (\mathbb{G}, *)$$

is a Lie group isomorphism. Choose a basis

$$B = (X_1^1, X_2^1, \ldots, X_{n_1}^1, X_1^2, \ldots, X_{n_2}^2, \ldots, X_1^r, \ldots, X_{n_r}^r)$$

of $\mathfrak{g}$ adapted to the stratification (we always consider a fixed stratification as in Definition 2.1) and consider the vector space isomorphism

$$T_B : \mathfrak{g} \longrightarrow \mathbb{R}^n$$

that sends B to the canonical basis of $\mathbb{R}^n$ with $n = \sum_{k=1}^r n_k$. We now define a new operation on $\mathbb{R}^n$. Given $x, y \in \mathbb{R}^n$ we set

$$x \star y = T_B(T_B^{-1}(x) \diamond T_B^{-1}(y))$$

which is a smooth group operation on $\mathbb{R}^n$, thus rendering $(\mathbb{R}^n, \star)$ a Lie group. Directly from the definition of $\star$ we see that L_B is a Lie group isomorphism between $(\mathfrak{g}, \diamond)$ and $(\mathbb{R}^n, \star)$, thus $T_B \circ \exp^{-1}$ is a Lie group isomorphism between $(\mathbb{G}, *)$ and $(\mathbb{R}^n, \star)$. $\qquad\square$

We describe now how to define a family of dilations $(\delta_\lambda)_{\lambda>0}$ on $(\mathbb{R}^n, \star)$ that are group homomorphisms. We relabel the basis $B = (X_1^1, \ldots, X_{n_r}^r)$ as $B = (X_1, \ldots, X_n)$ where instead of X_h^k we write X_j with $j = \sum_{i=1}^{k-1} n_i + h$ and use the notation $d_j = k$ for the weight of the coordinate j. We consequently denote points $x \in \mathbb{R}^n$ as $x = (x_1, \ldots, x_n)$. Thanks to the stratification we can define a family of dilations Δ_λ in $\mathfrak{g}$. An element $X \in \mathfrak{g}$ can be expressed as $X = \sum_{i=1}^n x_i X_i$ and we define

$$\Delta_\lambda(X) = \sum_{i=1}^n \lambda^{d_i} x_i X_i.$$

Thanks to Remark 2.2 the form of the Δ_λ is independent from the stratification and the choice of the adapted basis. It can be easily seen that the dilations Δ_λ are Lie algebra automorphisms of $\mathfrak{g}$, i.e.

$$\Delta_\lambda([X, Y]) = [\Delta_\lambda(X), \Delta_\lambda(Y)],$$

by observing that due to the stratification we have $[X_j, X_i] \in V_{d_j+d_i}$. Since the group operation $\diamond$ is defined as a finite sum of commutators of finite lenght and since Δ_λ commutes with the bracket operation we get that Δ_λ are also group automorphisms of $(\mathfrak{g}, \diamond)$. Now we define $\delta_\lambda : \mathbb{R}^n \longrightarrow \mathbb{R}^n$ as $\delta_\lambda = T_B \circ \Delta_\lambda \circ T_B^{-1}$. From the definition of $\star$ we can easily see that δ_λ are automorphisms of the Lie group $(\mathbb{R}^n, \star)$. Explicitly we have

$$\delta_\lambda(x_1, \ldots, x_n) = (\lambda^{d_1} x_1, \ldots, \lambda^{d_n} x_n).$$

Remark 2.4 If in the previous proof we choose a different basis B' of $\mathfrak{g}$ adapted to the stratification we obtain a different Lie group $(\mathbb{R}^n, \star')$ isomorphic to $(\mathbb{G}, *)$ via $\exp \circ L_{B'}^{-1}$. The Lie groups $(\mathbb{R}^n, \star)$ and $(\mathbb{R}^n, \star')$ are isomorphic, an isomorphism given by the change of basis $L_{B'} \circ L_B^{-1}$.

Remark 2.5 We point out some properties of the Lie group $(\mathbb{R}^n, \star)$, namely

- the identity element is 0,
- the inverse is given by $x^{-1} = -x$,
- the operation is of the form

$$(x \star y) = (x_1 + y_1, x_2 + y_2 + Q_2(x, y), \ldots, x_n + y_n + Q_n(x, y)) \qquad (2.4)$$

where

$$Q_j(\delta_\lambda(x), \delta_\lambda(x)) = \lambda^{d_j} Q_j(x, y)$$

and Q_j is a sum of monomials in x_k, y_k with $d_k < d_j$,
- a basis of the Lie algebra is given by vector fields X_j, $j = 1, \ldots, n$ of the form

$$X_j = \partial_j + \sum_{i=m_{d_j}+1}^n a_{i,j}(x)\partial_i \quad j = 1, \ldots, n-1$$

$$X_n = \partial_n \qquad (2.5)$$

with $a_{i,j}(\delta_\lambda(x)) = \lambda^{d_i - d_j} a_{i,j}(x)$.

Since we have seen that we can always identify a Carnot group with a Lie group on $\mathbb{R}^n$ we will work with the Carnot group $(\mathbb{R}^n, \star)$ without loss of generality.

The first layer $V_1 = \text{span}\{X_1^1, \ldots, X_{n_1}^1\}$ can be identified with a subspace of the tangent space at the identity $H_e\mathbb{G} \subset T_e\mathbb{G}$. Using the left translations $L_g : \mathbb{R}^n \longrightarrow \mathbb{R}^n$, $L_g(x) = g \star x$ we can define the horizontal subbundle $H\mathbb{G}$ of the tangent bundle $T\mathbb{G}$ whose fibers are given by

$$H_g\mathbb{G} = (L_g)_* H_e\mathbb{G} = \text{span}\{X_1^1(g), \ldots, X_{n_1}^1(g)\}.$$

We can introduce a sub-Riemannian metric providing a scalar product $\langle \, , \, \rangle_g$ on every fiber of the horizontal subbundle $H_g\mathbb{G}$ which varies smoothly with g. We choose it in such a way that $(X_1^1, \ldots, X_{n_1}^1)$ is orthonormal.

We will introduce a metric on $(\mathbb{R}^n, \star)$ different from the Euclidean one, the Carnot-Carathéodory distance, which is more natural in many ways. First we will provide some definitions and properties of Hörmander's vector fields.

Definition 2.5 A system of smooth vector fields $X = (X_1, \ldots, X_k)$ on an open subset $\Omega \subset \mathbb{R}^n$ is called a Hörmander's system if

$$\dim \operatorname{span}\{Y(x) \mid Y \in \operatorname{Lie}(X)\} = n \quad \text{for all } x \in \Omega \tag{2.6}$$

where $\operatorname{Lie}(X)$ is the Lie algebra generated by the set of vector fields X.

Remark 2.6 If $(\mathbb{R}^n, \star)$ is a Carnot group and $B = (X_1^1, X_2^1, \ldots, X_{n_1}^1, X_1^2, \ldots, X_{n_2}^2, \ldots, X_1^r, \ldots, X_{n_r}^r)$ is a basis of $\mathfrak{g}$ adapted to the stratification, then $X = (X_1^1, \ldots, X_{n_1}^1)$ is a Hörmander's system of vector fields.

Definition 2.6 Let $X = (X_1, \ldots, X_k)$ be a Hörmander's system of vector fields on $\Omega \subset \mathbb{R}^n$. An X-subunitary curve is an absolutely continuous curve $\Gamma : [0, T] \longrightarrow \Omega$ such that

$$\Gamma'(t) = \sum_{j=1}^{k} \alpha_j(t) X_j(\Gamma(t))$$

for some real valued functions α_j with $\sum_{j=1}^{k} \alpha_j(t)^2 \le 1$. We denote by $S(x, y)$ the set of all X-subunitary curves joining x and y.

We recall the following celebrated Theorem (a proof can be found in [1]).

Theorem 2.5 (Chow's accessibility Theorem) *Let Ω be an open connected subset of $\mathbb{R}^n$ and $X = (X_1, \ldots, X_k)$ a Hörmander's system of vector fields. Then for every couple of points $x, y \in \Omega$ there exists an X-subunitary curve $\Gamma \in S(x, y)$ joining x and y.*

Definition 2.7 If $\Gamma : [0, T] \longrightarrow (\mathbb{R}^n, \star)$ is an X-subunitary curve with respect to $X = (X_1^1, \ldots, X_{n_1}^1)$ we call it horizontal subunitary curve and we define its length as $l(\Gamma) = T$.

Definition 2.8 (*Carnot-Carathéodory distance*) We define the Carnot-Carathéodory distance on $(\mathbb{R}^n, \star)$ as:

$$d_{cc}(x, y) = \inf\{l(\Gamma) \mid \Gamma \in S(x, y)\}.$$

The previous definition is well posed thanks to Chow's accessibility Theorem 2.5.

When there is no possibility of confusion we will denote the Carnot-Carathéodory distance simply as d and the balls

$$B_r(x) = \{y \in \mathbb{R}^n \mid d(x, y) < r\}$$

will be with respect to the Carnot-Carathéodory distance unless otherwise stated.

We will provide some basic but very useful properties of the Carnot-Carathéodory distance, in particular the left-invariance and the homogeneity with respect to group dilations.

Proposition 2.1 *For all $x, y, g \in \mathbb{R}^n$ and $\lambda > 0$ we have*

$$d(L_g(x), L_g(y))) = d(x, y) \tag{2.7}$$
$$d(\delta_\lambda(x), \delta_\lambda(y)) = \lambda d(x, y) . \tag{2.8}$$

Proof It will suffice to prove that $\Gamma : [0, T] \longrightarrow \mathbb{R}^n$ belongs to $S(x, y)$ if and only if $L_g \circ \Gamma : [0, T] \longrightarrow \mathbb{G}$ belongs to $S(L_g(x), L_g(y))$. Since L_g is smooth then $L_g \circ \Gamma$ is absolutely continuous and it is obvious that $L_g \circ \Gamma(0) = L_g(x)$ and $L_g \circ \Gamma(T) = L_g(y)$. We can compute

$$(L_g \circ \Gamma)'(t) = (dL_g)(\Gamma(t)) \sum_{j=1}^{n_1} \alpha_j(t) X_j(\Gamma(t)) = \sum_{j=1}^{n_1} \alpha_j(t)(dL_g)(\Gamma(t)) X_j(\Gamma(t))$$

$$= \sum_{j=1}^{n_1} \alpha_j(t) X_j(L_g \circ \Gamma(t))$$

(2.9)

where in the last equality we have used the left-invariance of X_j.

For the second part it will suffice to prove that $\Gamma : [0, T] \longrightarrow \mathbb{R}^n$ is an X-subunitary curve if and only if $\Gamma_\lambda : [0, \lambda T] \longrightarrow \mathbb{G}$, $\Gamma_\lambda(t) = \delta_\lambda(\Gamma(\frac{t}{\lambda}))$ is an X-subunitary curve. We have

$$\Gamma'(t) = \sum_{j=1}^{n_1} \alpha_j(t) X_j(\Gamma(t)) = \sum_{j=1}^{n_1} \alpha_j(t) \sum_{i=1}^{n} a_{i,j}(\Gamma(t)) \partial_{x_i}. \qquad (2.10)$$

Now

$$\Gamma'_\lambda(t) = \sum_{i=1}^{n} \lambda^{d_i-1} \sum_{j=1}^{n_1} \alpha_j \left(\frac{t}{\lambda}\right) a_{i,j} \left(\Gamma\left(\frac{t}{\lambda}\right)\right) \partial_i = \sum_{i=1}^{n} \sum_{j=1}^{n_1} \alpha_j \left(\frac{t}{\lambda}\right) a_{i,j} \left(\delta_\lambda \circ \Gamma\left(\frac{t}{\lambda}\right)\right) \partial_i$$

$$= \sum_{j=1}^{n_1} \alpha_j \left(\frac{t}{\lambda}\right) X_j \left(\Gamma_\lambda(t)\right) \qquad (2.11)$$

since $a_{i,j}$ are homogeneous of degree $d_i - 1$ from Remark 2.5. $\qquad\qquad \square$

We refer to [1] for a proof of the following Theorem.

Theorem 2.6 *The Carnot-Carathéodory distance is continuous with respect to the Euclidean topology.*

Due to the anisotropy of the dilations δ_λ it is natural to consider norms that respect their particular form.

Definition 2.9 (*Homogeneous norm*) A homogeneous (symmetric) norm on $(\mathbb{R}^n, \star)$ is a continuous function (with respect to the Euclidean topology)

$$| \cdot |_G : \mathbb{R}^n \longrightarrow [0, +\infty[$$

such that

1. $|x|_G = 0$ if and only if $x = 0$,
2. $|\delta_\lambda(x)|_G = \lambda|x|_G$ for all $x \in \mathbb{R}^n$, $\lambda > 0$,
3. $|x^{-1}|_G = |x|_G$ for all $x \in \mathbb{R}^n$.

Remark 2.7 The Carnot-Carathéodory distance defines a homogeneous norm in the following way

$$|x|_{CC} = d_{CC}(x, 0).$$

Lemma 2.1 *Let* $|\cdot|_G$ *be a homogeneous norm on* $(\mathbb{R}^n, \star)$. *Then the ball*

$$B_G = \{y \in \mathbb{R}^n \mid |y|_G = 1\}$$

is compact (with respect to the Euclidean topology).

Proof Consider the function $|\cdot|_S : \mathbb{R}^n \longrightarrow [0, \infty[$ given by

$$|x|_S = \sum_{i=1}^{n} |x_i|^{\frac{1}{d_i}} \tag{2.12}$$

where d_i is the weight of the coordinate i. In this way $|\cdot|_S$ is a homogeneous norm. We can readily see that the set $B_S = \{y \in \mathbb{R}^n \mid |y|_S = 1\}$ is compact with respect to the Euclidean topology because it is closed and bounded. Moreover it doesn't contain 0 thus the function $|\cdot|_G$ attains a minimum $\nu > 0$ on this set

$$\nu \leq |y|_G \quad \text{for all } y \in B_S.$$

By the homogeneity property we have

$$\nu \leq \left|\delta_{\frac{1}{|x|_S}}(x)\right|_G = \frac{|x|_G}{|x|_S} \quad \text{for all } x \in \mathbb{R}^n.$$

This implies that

$$B_G = \{y \in \mathbb{R}^n \mid |y|_G = 1\} \subseteq \{y \in \mathbb{R}^n \mid |y|_S \leq 1\}$$

therefore it is compact. $\qquad\square$

Proposition 2.2 *Every homogeneous norm* $|\cdot|_G$ *on* $(\mathbb{R}^n, \star)$ *is equivalent to* $|\cdot|_{CC}$.

Proof Since d_{CC} is continuous by Theorem 2.6 and $B = \{y \in \mathbb{R}^n \mid |y|_G = 1\}$ is compact by Lemma 2.1 there exist two positive constants m and M such that

$$m \leq d_{CC}(y, 0) \leq M \quad \text{for all } y \in B.$$

Thanks to the homogeneity of the dilations (2.8) we have

$$m \leq \frac{d_{CC}(x,0)}{|x|_G} = d_{CC}\left(\delta_{\frac{1}{|x|_G}}(x),0\right) \leq M \quad \text{for all } x \in G$$

which gives

$$m|x|_G \leq |x|_{CC} \leq M|x|_G \quad \text{for all } x \in G. \qquad \square$$

Proposition 2.3 *The Lebesgue measure* dx *is the Haar measure of the group* $(\mathbb{R}^n, \star)$, *namely it is left- and right-invariant with respect to group translations. Moreover we have*

$$dx\,(\delta_\lambda(A)) = \lambda^Q dx(A) \qquad (2.13)$$

for every measurable set A.

Proof We have to prove $dx(L_g(A)) = dx(A) = dx(R_g(A))$ for every measurable set A. By a change of variables we have

$$dx(L_g(A)) = \int_{L_g(A)} dx = \int_A |\det J_{L_g}(x)|\ dx = \int_A dx = dx(A)$$

where we have used the fact that thanks to (2.4) the Jacobian matrix has the lower triangular form

$$J_{L_g}(x) = \begin{pmatrix} 1 & 0 & 0 & \cdots & 0 \\ a_{1,2}(x) & 1 & 0 & \cdots & 0 \\ a_{1,3}(x) & a_{2,3}(x) & 1 & \cdots & 0 \\ \vdots & \ddots & \ddots & \ddots & \vdots \\ a_{1,n}(x) & a_{2,n}(x) & a_{3,n}(x) & \cdots & 1 \end{pmatrix}.$$

Analogously we get the result for the right translations.

For the second part observe that the Jacobian matrix of the dilation δ_λ is a diagonal matrix

$$J_{\delta_\lambda} = \text{diag}(\lambda, \ldots, \lambda, \lambda^2, \ldots, \lambda^2, \ldots, \lambda^r, \ldots, \lambda^r)$$

where λ^i appears a number $\dim(V_i)$ times so that the determinant is

$$\lambda^{\sum_{i=1}^r i \dim(V_i)} = \lambda^Q. \qquad \square$$

Definition 2.10 In the notations of the proof of Theorem 2.4 we can define a left- and right-invariant measure μ on the group $(\mathbb{G}, \ast)$ by means of the Lie group isomorphism $T_B \circ \exp^{-1}$. We define

$$\mu = (T_B \circ \exp^{-1})^* dx$$

which means $\mu(A) = dx\left((T_B \circ \exp^{-1})(A)\right)$. It is easy to show that μ is actually left- and right-invariant. Indeed if we call τ_g the left translation on $(\mathbb{G}, \ast)$ we have

$$\mu\left(\tau_g(A)\right) = dx\left((T_B \circ \exp^{-1})(\tau_g(A))\right) = dx\left(L_{T_B \circ \exp^{-1}(g)} \circ T_B \circ \exp^{-1}(A)\right)$$
$$= dx\left(T_B \circ \exp^{-1}(A)\right) = \mu(A)$$

where we have used the Lie group isomorphism property and the left invariance of the measure dx.

Remark 2.8 The measure μ does not depend on the choice of the basis used for the identification with $\mathbb{R}^n$. In fact using the same notations as in the proof of Theorem 2.4 and Remark 2.4, choosing two different basis adapted to the stratification B and B' we have two Lie groups isomorphisms

$$T_B \circ \exp^{-1} : (\mathbb{G}, *) \longrightarrow (\mathbb{R}^n, \star)$$
$$T_{B'} \circ \exp^{-1} : (\mathbb{G}, *) \longrightarrow (\mathbb{R}^n, \star').$$

Following the construction of Definition 2.10 we can define on $(\mathbb{G}, *)$ two measures $\mu = (T_B \circ \exp^{-1})^*dx$ and $\mu' = (T_{B'} \circ \exp^{-1})^*dx$. From Remark 2.4 the Lie groups $(\mathbb{R}^n, \star)$ and $(\mathbb{R}^n, \star')$ are isomorphic via the linear transformation $T_{B'} \circ T_B^{-1}$, whose Jacobian is 1. Therefore we have

$$\mu'(A) = dx\left(T_{B'} \circ \exp^{-1}(A)\right) = dx\left(T_{B'} \circ T_B^{-1} \circ T_B \circ \exp^{-1}(A)\right)$$
$$= \int_{T_B \circ \exp^{-1}(A)} \left|\det J_{T_{B'} \circ T_B^{-1}}\right| dx = dx\left(T_B \circ \exp^{-1}(A)\right) = \mu(A)$$

so the two measures μ and μ' coincide.

2.2 The Heisenberg Group

In this section we will introduce an explicit example of Carnot group which will be the setting of the rest of this monograph.

The first Heisenberg group $\mathbb{H}$ is a step 2 Carnot group which can be identified with the Lie group $(\mathbb{R}^3, \star)$. Indicating points $x, y \in \mathbb{H}$ by $x = (x_1, x_2, z)$ and $y = (y_1, y_2, s)$ the group operation is

$$x \star y = (x_1, x_2, z) \star (y_1, y_2, s) = \left(x_1 + y_1, x_2 + y_2, z + s + \frac{1}{2}(x_1 y_2 - x_2 y_1)\right).$$
$$(2.14)$$

A basis of left-invariant vector fields for the associated Lie algebra $\mathfrak{h}$ is given by

$$X_1 = \partial_{x_1} - \frac{x_2}{2}\partial_z,$$
$$X_2 = \partial_{x_2} + \frac{x_1}{2}\partial_z,$$

$$T = \partial_z. \tag{2.15}$$

The only non vanishing commutator is $[X_1, X_2] = T$. We obtain the stratification of the Lie algebra $\mathfrak{h} = \mathfrak{h}_1 \oplus \mathfrak{h}_2$, where

$$\mathfrak{h}_1 = \text{span}\{X_1, X_2\} \tag{2.16}$$
$$\mathfrak{h}_2 = \text{span}\{T\} \tag{2.17}$$

with $[\mathfrak{h}_1, \mathfrak{h}_1] = \mathfrak{h}_2$.

Let Ω be an open subset of $\mathbb{H}$ and consider a function $u : \Omega \longrightarrow \mathbb{R}$. We will indicate by $\nabla_{\mathbb{H}} u = (X_1 u, X_2 u)$ the horizontal gradient of u and by $\nabla_{\mathbb{H}}^2 u = (X_i X_j u)_{i,j=1,2}$ the horizontal hessian of u. If $X = a_1 X_1 + a_2 X_2$ is a horizontal vector field we denote by $\text{div}_{\mathbb{H}} X = X_1 a_1 + X_2 a_2$ its horizontal divergence.

Let $x = (x_1, x_2, z) \in \mathbb{H}$, $X \in \mathfrak{h}$ and $t \in \mathbb{R}$. Expressing $X = \xi_1 X_1 + \xi_2 X_2 + \xi_3 T$ we can compute explicitly the integral curve of X at time t starting at x:

$$e^{tX} x = \left(\xi_1 t + x_1, \xi_2 t + x_2, \frac{1}{2}(\xi_2 x_1 - \xi_1 x_2)t + \xi_3 \right).$$

Considering e^{tX} as a function from $\mathbb{R}^3$ to $\mathbb{R}^3$ its Jacobian matrix is

$$\begin{pmatrix} 1 & 0 & 0 \\ 0 & 1 & 0 \\ \frac{1}{2}\xi_2 & -\frac{1}{2}\xi_1 & 1 \end{pmatrix} \tag{2.18}$$

and we can see that the determinant is 1, as already showed in the previous section. The Baker–Campbell–Hausdorff formula is

$$e^X e^Y = e^{X+Y+\frac{1}{2}[X,Y]} \tag{2.19}$$

and we have the well known relation

$$e^{-hX} e^{-hY} e^{hX} e^{hY} x = e^{h^2[X,Y]} x. \tag{2.20}$$

In what follows all the balls will be with respect to the Carnot-Carathéodory distance, unless otherwise stated. We will also use a particular homogenous norm, the Korányi norm

$$|x|_K = \left((x_1^2 + x_2^2)^2 + z^2 \right)^{\frac{1}{4}}$$

which is equivalent to the Carnot-Carathéodory norm $|x|_c = d(x, 0)$ thanks to Proposition 2.2. The dilations $(\delta_\lambda)_{\lambda > 0}$ become in this case

$$\delta_\lambda(x_1, x_2, z) = (\lambda x_1, \lambda x_2, \lambda^2 z).$$

As described in the previous section the Lebesgue measure in $\mathbb{R}^3$ is the Haar measure of the group and we will denote the measure of a set A by $|A|$. The homogeneous dimension of $\mathbb{H}$ is $Q = 4$.

The horizontal Sobolev space $HW^{1,p}(\Omega)$ is the space of $L^p(\Omega)$ functions u whose first horizontal derivatives $X_1 u$ and $X_2 u$ are in $L^p(\Omega)$. It is a Banach space if endowed with the Sobolev norm

$$\|u\|_{HW^{1,p}(\Omega)} = \|u\|_{L^p(\Omega)} + \|\nabla_{\mathbb{H}} u\|_{L^p(\Omega)}$$

and it is reflexive if $1 < p < \infty$. Analogously to the Euclidean case we can approximate $HW^{1,p}$ functions with smooth functions (see [3] and the references therein).

Theorem 2.7 *Let Ω be an open subset of $\mathbb{R}^n$ and $1 \leq p < \infty$. Then*

$$C^\infty(\Omega) \cap HW^{1,p}(\Omega) \text{ is dense in } HW^{1,p}(\Omega).$$

We also define the space $HW_0^{1,p}(\Omega)$ as the closure of $C_0^\infty(\Omega)$ in $HW^{1,p}(\Omega)$ with respect to the norm $\|\cdot\|_{HW^{1,p}(\Omega)}$.

We denote the average of a function f over a set B by

$$f_B = \fint_B f(x)\mathrm{d}x = \frac{1}{|B|} \int_B f(x)\mathrm{d}x.$$

We will need the following Sobolev inequality for the Heisenberg group (valid in more general settings, see [4]).

Theorem 2.8 *Let $B_r \subset \mathbb{H}$ and $1 < q < Q$. Then*

$$\left(\fint_{B_r} |u|^{\frac{Qq}{Q-q}} \right)^{\frac{Q-q}{Qq}} \leq C_q r \left(\fint_{B_r} |\nabla_{\mathbb{H}} u|^q \right)^{\frac{1}{q}} \tag{2.21}$$

for all $u \in HW_0^{1,q}(B_r)$.

We will need also a Rellich type Theorem and a subelliptic version of the Poincaré inequality (see [3]).

Theorem 2.9 *Let Ω be a bounded open subset of $\mathbb{H}$ and $1 < q < Q$. Then $HW_0^{1,q}(\Omega)$ is compactly embedded in $L^{\frac{qQ}{Q-q}}(\Omega)$.*

Theorem 2.10 *Let $1 < q \leq p < \infty$ and $u \in HW^{1,q}(B_r)$. Then*

$$\left(\fint_{B_r} |u - u_{B_r}|^p \right)^{\frac{1}{p}} \leq Cr \left(\fint_{B_r} |\nabla_{\mathbb{H}} u|^q \right)^{\frac{1}{q}}. \tag{2.22}$$

2.3 Difference Quotients

Difference quotients are discrete versions of derivatives, and they are a useful tool in proving differentiability results.

Definition 2.11 (*Difference Quotients*) Let $u : \Omega \longrightarrow \mathbb{R}$ be a function and Z be a left-invariant vector field. Fix $\alpha \in]0, 1]$ and $h \in \mathbb{R}$. We define for those h for which $e^{hZ}x, e^{-hZ}x \in \Omega$ the difference quotients

$$\Delta^{\alpha}_{Z,h} u(x) = \frac{u(e^{hZ}x) - u(x)}{h^{\alpha}},$$

$$\Delta_{Z,h} u(x) = \Delta^{1}_{Z,h} u(x),$$

$$\Delta^{2,\alpha}_{Z,h} u(x) = \frac{u(e^{hZ}x) + u(e^{-hZ}x) - 2u(x)}{h^{\alpha}}.$$

Remark 2.9 Observe that directly from the definitions it follows that

$$\Delta^{\alpha_1}_{Z,-h} \left(\Delta^{\alpha_2}_{Z,h} u(x) \right) = \Delta^{\alpha_2}_{Z,h} \left(\Delta^{\alpha_1}_{Z,-h} u(x) \right) = \Delta^{2,\alpha_1+\alpha_2}_{Z,h} u(x).$$

Note that vector fields and difference quotients don't commute in general as in the Euclidean case.

Lemma 2.2 *Let X and $Y \in \mathfrak{h}$, $h \in \mathbb{R}$ and $\alpha > 0$. Then*

$$X \left(\Delta^{\alpha}_{Y,h} u(x) \right) = \Delta^{\alpha}_{Y,h} \left(Xu(x) \right) + [X, Y] u(e^{hY}x). \tag{2.23}$$

For difference quotients we have a discrete integration by parts formula.

Proposition 2.4 *Let $u : \Omega \to \mathbb{R}$, $\alpha \in]0, 1]$, $\varphi \in C^{\infty}_0(\Omega)$ and Z a left-invariant vector field. Then*

$$\int_{\Omega} \Delta^{\alpha}_{Z,h} u \, \varphi \, dx = \int_{\Omega} u \, \Delta^{\alpha}_{Z,-h} \varphi \, dx$$

for $h > 0$ such that $e^{hZ}(\mathrm{supp}\varphi)$ and $e^{-hZ}(\mathrm{supp}\varphi) \subset \Omega$.

Proof Let $K = \mathrm{supp}\varphi$. We have

$$\int_{\Omega} \Delta^{\alpha}_{Z,h} u \, \varphi \, dx = \frac{1}{h^{\alpha}} \left(\int_{K} u(e^{hZ}x) \, \varphi(x) \, dx - \int_{\Omega} u(x) \, \varphi(x) \, dx \right). \tag{2.24}$$

Now changing variables in the first integral we get

$$\int_{K} u(e^{hZ}x) \, \varphi(x) \, dx = \int_{\Omega} u(x) \, \varphi(e^{-hZ}x) \, dx$$

because the determinant of the Jacobian matrix of the change of variables is 1 by (2.18), and h is such that $e^{hZ}(K) \subset \Omega$.

In this way (2.24) becomes

$$\int_\Omega \Delta^\alpha_{Z,h} u \, \varphi \, dx = \int_\Omega u(x) \frac{\varphi(e^{-hZ}x) - \varphi(x)}{h^\alpha} dx$$

which is the desired result. □

Difference quotients can be used to give a characterization of $HW^{1,p}(\Omega)$ functions, analogous to the Euclidean one. More precisely we have the following Theorem.

Theorem 2.11 *Let $u \in HW^{1,p}_{loc}(\Omega)$, $1 < p < \infty$ and $K \subset \Omega$ a compact subset. Suppose there exist constants $C, \varepsilon > 0$ such that $e^{hZ}(K) \subset \Omega$ for all $0 < |h| < \varepsilon$ and*

$$\sup_{0<|h|<\varepsilon} \left\| \Delta_{Z,h} u \right\|_{L^p(K)} \le C.$$

Then $Zu \in L^p(K)$ and $\|Zu\|_{L^p(K)} \le C$.
  Conversely if $Zu \in L^p_{loc}(\Omega)$ then

$$\sup_{0<|h|<\varepsilon} \left\| \Delta_{Z,h} u \right\|_{L^p(K)} \le \|Zu\|_{L^p(K')}$$

for $K \subset K' \subset \Omega$.

Proof We have to prove that there exists $g \in L^p_{loc}(\Omega)$ such that

$$\int_\Omega f Z^* \varphi \, dx = \int_\Omega g\varphi \, dx.$$

Take a sequence $h_n \to 0$, then $\left\| \Delta_{Z,h_n} u \right\|_{L^p(K)} \le C$ when n is big enough. Since $1 < p < \infty$ there exists a function $g \in L^p(K)$ such that (possibly passing to a subsequence which we continue to denote by h_n) we have $\Delta_{Z,h_n} u \xrightarrow{w} g$ in $L^p(K)$. Let $\varphi \in C^\infty_0(K)$. By the discrete integration by parts formula in Proposition 2.4 we have

$$\int_K \Delta_{Z,h_n} u \, \varphi \, dx = \int_\Omega u \Delta_{Z,-h_n} \varphi \, dx \longrightarrow -\int_\Omega u \, Z\varphi \, dx = \int_\Omega u \, Z^* \varphi \, dx$$

thanks to Lebesgue's dominated convergence theorem and because in the Heisenberg group $Z^* = -Z$.
  On the other hand by weak convergence we get

$$\int_K \Delta_{Z,h_n} u \, \varphi \, dx \longrightarrow \int_\Omega g \, \varphi \, dx$$

and this concludes the first part of the proposition.

For the second part observe that

$$u(e^{hZ}x) - u(x) = \int_0^1 \frac{d}{dt} u(e^{thZ}x)\, dt = \int_0^1 (hZ)\, u(e^{thZ}x)\, dt.$$

Now by Jensen's inequality

$$\int_K \left| \frac{u(e^{hZ}x) - u(x)}{h} \right|^p dx = \int_K \left| \int_0^1 Zu(e^{thZ}x)\, dt \right|^p dx$$

$$\leq \int_K \int_0^1 \left| Zu(e^{thZ}x) \right|^p dt\, dx = \int_0^1 \int_K \left| Zu(e^{thZ}x) \right|^p dx\, dt.$$

Now performing a change of variables we get

$$\int_0^1 \int_K \left| Zu(e^{thZ}x) \right|^p dx\, dt = \int_0^1 \int_{e^{thZ}(K)} |Zu(x)|^p dx\, dt \leq \|Zu\|_{L^p(K')}$$

for h such that $e^{hZ}(K) \subset K'$.   □

The next result is due to Domokos [2]. It allows us to control the first order difference quotients if we have a bound on the second order difference quotients.

Theorem 2.12 *Let $u \in L^p(\mathbb{H})$, $\alpha > 0$, $\varepsilon > 0$, $M > 0$ and Z a left invariant vector field such that*

$$\sup_{0 < |h| < \varepsilon} \left\| \Delta_{Z,h}^{2,\alpha} u \right\|_{L^p(\mathbb{H})} \leq M. \tag{2.25}$$

Then there exists a constant $C > 0$ such that

$$\sup_{0 < |h| < \bar{\varepsilon}} \left\| \Delta_{Z,h}^{\beta} u \right\|_{L^p(\mathbb{H})} \leq C(\|u\|_{L^p(\mathbb{H})} + M), \tag{2.26}$$

where

$$\begin{cases} \beta = \alpha & \text{if } 0 < \alpha < 1 \\ \beta \in\,]0, 1[& \text{if } \alpha = 1 \\ \beta = 1 & \text{if } \alpha > 1. \end{cases} \tag{2.27}$$

Proof Denote by $D_h u(x) = h\Delta_{Z,h} u(x) = u(e^{hZ}x) - u(x)$. By (2.25)

$$\left\| u(e^{hZ}x) + u(e^{-hZ}x) - 2u(x) \right\|_{L^p} \leq M|h|^\alpha \tag{2.28}$$

holds for all $h < \varepsilon$. Use the sequence $h_n = \frac{h}{2^n}$ to get

$$\left\| u(e^{h_n Z}x) + u(e^{-h_n Z}x) - 2u(x) \right\|_{L^p} \leq M|h_n|^\alpha. \tag{2.29}$$

Now changing variables, since the Jacobian determinant of the transformation is 1 by (2.18), we get

$$\left\| u(e^{h_{n-1}Z}x) + u(x) - 2u(e^{h_n Z}x) \right\|_{L^p} \le M|h_n|^{\alpha} \tag{2.30}$$

which can be written as

$$\left\| u(e^{h_{n-1}Z}x) - u(x) - 2\left(u(e^{h_n Z}x) - u(x) \right) \right\|_{L^p} \le M|h_n|^{\alpha}. \tag{2.31}$$

Now multiplying by 2^{n-1} we get

$$\left\| 2^{n-1} D_{h_{n-1}} u - 2^n D_{h_n} u \right\|_{L^p} \le \frac{M}{2^{n\alpha}} 2^{n-1} |h|^{\alpha} \tag{2.32}$$

and by the triangle inequality

$$\left\| D_h u - 2^n D_{h_n} u \right\|_{L^p} \le \sum_{j=0}^{n-1} \left\| 2^j D_{h_j} u - 2^{j+1} D_{h_{j+1}} u \right\|_{L^p}$$

$$\le \frac{M}{2} |h|^{\alpha} \sum_{j=0}^{n-1} 2^{j(1-\alpha)}. \tag{2.33}$$

If $0 < \alpha < 1$ we get

$$\left\| D_h u - 2^n D_{h_n} u \right\|_{L^p} \le \frac{M}{2} |h|^{\alpha} \frac{2^{n(1-\alpha)}}{2^{1-\alpha} - 1} \tag{2.34}$$

so again by the triangle inequality

$$\left\| D_{h_n} u \right\|_{L^p} \le |h|^{\alpha} \frac{M}{2^{n+1}} \frac{2^{n(1-\alpha)}}{2^{1-\alpha} - 1} + \frac{\|D_h u\|_{L^p}}{2^n} \le |h|^{\alpha} \frac{M}{2^{n+1}} \frac{2^{n(1-\alpha)}}{2^{1-\alpha} - 1} + \frac{1}{2^{n-1}} \|u\|_{L^p}. \tag{2.35}$$

Now fix $\delta < \varepsilon/2$, and consider $0 < k < a$. There exist $h \in [\delta/2, \delta]$ and $n \in \mathbb{N}$ such that $k = \frac{h}{2^n}$. In this way (2.35) becomes

$$\|D_k u\|_{L^p} \le \frac{|k|^{\alpha}}{2^{n\alpha}} \frac{M}{2} \frac{2^{-n\alpha)}}{2^{1-\alpha} - 1} + \frac{2k}{h} \|u\|_{L^p}. \tag{2.36}$$

Now divinding by $|k|^{\alpha}$ we get

$$\left\| \Delta_{Z,k}^{\alpha} u \right\|_{L^p} \le \frac{M}{2^{1-\alpha}} + C\frac{4}{\delta} \|u\|_{L^p}. \tag{2.37}$$

If $\alpha = 1$ then (2.33) implies

$$\left\| D_h u - 2^n D_{h_n} u \right\|_{L^p} \le \frac{M}{2} |h| n \qquad (2.38)$$

so

$$\left\| D_{h_n} u \right\|_{L^p} \le |h| \frac{M}{2^{n+1}} n + \frac{1}{2^{n-1}} \|u\|_{L^p} . \qquad (2.39)$$

Choosing δ, k as we did before we get

$$\|D_k u\|_{L^p} \le |k| \frac{M}{2} \log_2 \frac{\delta}{k} + \frac{4|k|}{\delta} \|u\|_{L^p} , \qquad (2.40)$$

and dividing by $|k|^\beta$, $0 < \beta < 1$ we get

$$\left\| \Delta_{Z,k}^\beta u \right\|_{L^p} \le \frac{M}{2} |k|^{1-\beta} \log_2 \frac{\delta}{k} + \frac{4|k|^{1-\beta}}{\delta} \|u\|_{L^p} \qquad (2.41)$$

and we can conclude since $|k|^{1-\beta} \log_2 \frac{\delta}{k}$ and $|k|^{1-\beta}$ are bounded near the origin.
If $\alpha > 1$ then (2.33) implies

$$\left\| D_h u - 2^n D_{h_n} u \right\|_{L^p} \le \frac{M}{2} |h|^\alpha \frac{1}{1 - 2^{1-\alpha}} \qquad (2.42)$$

so analogously as we did before

$$\|D_k u\|_{L^p} \le \frac{M}{2} |k| |\delta|^{\alpha-1} \frac{1}{1 - 2^{1-\alpha}} + \frac{4|k|}{\delta} \|u\|_{L^p} \qquad (2.43)$$

and dividing by $|k|$ we get the result. $\square$

Theorem 2.13 *Let $u \in HW^{1,p}(\Omega)$, $1 < p < \infty$, B_r and B_R two balls such that $B_r \subset B_R \subset \Omega$. Then there exists $\varepsilon > 0$ such that*

$$\sup_{0 < |h| < \varepsilon} \left\| \Delta_{T,h}^{\frac{1}{2}} u \right\|_{L^p(B_r)} \le C_p \|\nabla_{\mathbb{H}} u\|_{L^p(B_R)} \qquad (2.44)$$

where C_p is a positive constant depending only on p.

Proof By (2.20) we have

$$\begin{aligned}
u(e^{h^2 T} x) - u(x) &= u(e^{-hX_1} e^{-hX_2} e^{hX_1} e^{hX_2} x) - u(e^{-hX_2} e^{hX_1} e^{hX_2} x) \\
&\quad + u(e^{-hX_2} e^{hX_1} e^{hX_2} x) - u(e^{hX_1} e^{hX_2} x) \\
&\quad + u(e^{hX_1} e^{hX_2} x) - u(e^{hX_2} x) \\
&\quad + u(e^{hX_2} x) - u(x).
\end{aligned} \qquad (2.45)$$

Raising to the power p and integrating both sides we get

$$\int_{B_r} |u(e^{h^2 T}x) - u(x)|^p dx \leq C_p \int_{B_r} |u(e^{-hX_1}e^{-hX_2}e^{hX_1}e^{hX_2}x) - u(e^{-hX_2}e^{hX_1}e^{hX_2}x)|^p dx$$

$$+ C_p \int_{B_r} |u(e^{-hX_2}e^{hX_1}e^{hX_2}x) - u(e^{hX_1}e^{hX_2}x)|^p dx$$

$$+ C_p \int_{B_r} |u(e^{hX_1}e^{hX_2}x) - u(e^{hX_2}x)|^p dx$$

$$+ C_p \int_{B_r} |u(e^{hX_2}x) - u(x)|^p dx. \tag{2.46}$$

Dividing by h and changing variables in each integral, by (2.18) and provided h is small enough

$$\left\| \Delta_{T,h^2}^{\frac{1}{2}} u \right\|_{L^p(B_r)} \leq C_p \left\| \Delta_{X_1,-h} u \right\|_{L^p(B_{r'})} + C_p \left\| \Delta_{X_2,-h} u \right\|_{L^p(B_{r'})}$$

$$+ C_p \left\| \Delta_{X_1,h} u \right\|_{L^p(B_{r'})} + C_p \left\| \Delta_{X_2,h} u \right\|_{L^p(B_{r'})}$$

$$\leq C_p \left\| \nabla_{\mathbb{H}} u \right\|_{L^p(B_R)} \tag{2.47}$$

where $r < r' < R$ and in last inequality we have used Theorem 2.11. □

2.4 Morrey and Campanato Spaces

Definition 2.12 Let $0 < \alpha \leq 1$. We define the Hölder class with respect to the Carnot-Carathéodory distance

$$\Gamma^\alpha(\Omega) = \left\{ u : \Omega \to \mathbb{R} \,\middle|\, \sup_{\substack{x,y \in \Omega \\ x \neq y}} \frac{|u(x) - u(y)|}{d(x,y)^\alpha} < \infty \right\}.$$

For $k \in \mathbb{N}$ we define

$$\Gamma^{k,\alpha}(\Omega) = \left\{ u : \Omega \to \mathbb{R} \,\middle|\, X^a u \in \Gamma^\alpha, \ a \text{ multiindex such that } |a| = k \right\}.$$

We will introduce some function spaces which are very useful to prove regularity results. Let $\Omega \subset \mathbb{H}$ be an open set, $0 < \lambda < 1$, $1 \leq p < \infty$ and $u \in L^p_{loc}(\Omega)$. Denote by $\Omega_r(x) = \Omega \cap B_r(x)$.

Definition 2.13 (*Morrey Space*) The function u is in the Morrey Space $M^{p,\lambda}(\Omega)$ if

$$\sup_{\substack{r > 0 \\ x \in \Omega}} \frac{1}{r^{p(\lambda-1)}} \fint_{\Omega_r(x)} |u(z)|^p dz < \infty. \tag{2.48}$$

Definition 2.14 (*Campanato Space*) The function u is in the Campanato Space $\mathcal{L}^{p,\lambda}(\Omega)$ if

$$\sup_{\substack{r>0 \\ x\in\Omega}} \frac{1}{r^{p\lambda}} \fint_{\Omega_r(x)} |u(z) - u_{\Omega_r(x)}(z)|^p \mathrm{d}z < \infty. \tag{2.49}$$

Remark 2.10 If $\nabla_{\mathbb{H}} u \in M^{p,\lambda}(B_{2r})$ then by the subelliptic version of Poincaré inequality (2.10) we have that $u \in \mathcal{L}^{p,\lambda}(B_r)$.

Definition 2.15 An open set $\Omega \subset \mathbb{H}$ satisfies the A-property if there exists a positive constant A such that

$$|\Omega_r(x)| \geq A|B_r(x)| \tag{2.50}$$

for all $x \in \overline{\Omega}$ and sufficiently small $r > 0$.

Lemma 2.3 *If Ω satisfies the A-property then $\mathcal{L}^{p,\lambda}(\Omega) \subset \Gamma^{\lambda}(\Omega)$.*

Proof Let $u \in \mathcal{L}^{p,\lambda}(\Omega)$ and fix $x, y \in \Omega$. Call $r = d(x, y)$. We have to prove that $|u(x) - u(y)| \leq Cr^{\lambda}$. We have

$$|u(x) - u(y)| \leq |u(x) - u_{B_{2r}(x)}| + |u_{B_{2r}(x)} - u_{B_{2r}(y)}| + |u_{B_{2r}(y)} - u(y)|. \tag{2.51}$$

Now observe that $B_r(x) \cup B_r(y) \subset B_{2r}(x) \cap B_{2r}(y)$ and

$$|\Omega_r(x) \cup \Omega_r(y)| \, |u_{B_{2r}(x)} - u_{B_{2r}(y)}| \leq \int_{\Omega_{2r}(x)\cap\Omega_{2r}(y)} |u_{B_{2r}(x)} - u_{B_{2r}(y)}| \, \mathrm{d}z$$

$$\leq \int_{\Omega_{2r}(x)} |u(z) - u_{B_{2r}(x)}| \, \mathrm{d}z$$

$$+ \int_{\Omega_{2r}(y)} |u(z) - u_{B_{2r}(y)}| \, \mathrm{d}z \tag{2.52}$$

therefore

$$|u_{B_{2r}(x)} - u_{B_{2r}(y)}| \leq \frac{1}{|\Omega_r(x)|} \int_{\Omega_{2r}(x)} |u(z) - u_{B_{2r}(x)}| \, \mathrm{d}z + \frac{1}{|\Omega_r(y)|} \int_{\Omega_{2r}(y)} |u(z) - u_{B_{2r}(y)}| \, \mathrm{d}z$$

$$= \frac{|\Omega_{2r}(x)|}{|\Omega_r(x)|} \fint_{\Omega_{2r}(x)} |u(z) - u_{B_{2r}(x)}| \, \mathrm{d}z + \frac{|\Omega_{2r}(y)|}{|\Omega_r(y)|} \fint_{\Omega_{2r}(y)} |u(z) - u_{B_{2r}(y)}| \, \mathrm{d}z$$

$$\leq C \left(\fint_{\Omega_{2r}(x)} |u(z) - u_{B_{2r}(x)}| \, \mathrm{d}z + \fint_{\Omega_{2r}(y)} |u(z) - u_{B_{2r}(y)}| \, \mathrm{d}z \right)$$

$$\leq C \left(\left(\fint_{\Omega_{2r}(x)} |u(z) - u_{B_{2r}(x)}|^p \mathrm{d}z \right)^{\frac{1}{p}} + \left(\fint_{\Omega_{2r}(y)} |u(z) - u_{B_{2r}(y)}|^p \mathrm{d}z \right)^{\frac{1}{p}} \right)$$

$$\leq Cr^{\lambda} \tag{2.53}$$

where we have used the scaling property of the measure (2.13), Hölder's inequality and the definition of Campanato space (2.49).

Now we will estimate the remaining terms. Considering $0 < \rho < r$ we have

$$|u_{B_\rho(x)} - u_{B_r(x)}|^p \le C_p \left(|u_{B_\rho(x)} - u(z)|^p + |u(z) - u_{B_r(x)}|^p \right). \tag{2.54}$$

Now averaging over $\Omega_\rho(x)$ we get

$$
\begin{aligned}
|u_{B_\rho(x)} - u_{B_r(x)}|^p &\le C_p \left(\fint_{\Omega_\rho(x)} |u_{B_\rho(x)} - u(z)|^p dz + \fint_{\Omega_\rho(x)} |u(z) - u_{B_r(x)}|^p dz \right) \\
&\le C_p \left(\rho^{\lambda p} + \frac{|\Omega_r(x)|}{|\Omega_\rho(x)|} r^{\lambda p} \right).
\end{aligned}
\tag{2.55}
$$

Now from the scaling property of the measure (2.13) we have

$$|B_r(x)| \le C \left(\frac{r}{\rho} \right)^Q |B_\rho(x)| \tag{2.56}$$

and since Ω satisfies the A-property we get

$$\frac{|\Omega_r(x)|}{|\Omega_\rho(x)|} \le \frac{|\Omega \cap B_r(x)|}{A\,|B_\rho(x)|} \le C \left(\frac{r}{\rho} \right)^Q. \tag{2.57}$$

Now (2.55) becomes

$$|u_{B_\rho(x)} - u_{B_r(x)}|^p \le C_p r^{\lambda p} \left(\left(\frac{\rho}{r} \right)^{\lambda p} + \left(\frac{r}{\rho} \right)^Q \right) \le C_p r^{\lambda p} \left(\frac{r}{\rho} \right)^Q. \tag{2.58}$$

Consider a sequence of radii $r_i = \frac{2r}{2^i}$ and let $u_i = u_{B_{r_i}(x)}$. If $0 < i < j$ we have

$$|u_j - u_i| \le \sum_{k=i}^{j-1} |u_{k+1} - u_k| \le C \sum_{k=i}^{j-1} r_i^\lambda \left(\frac{r_i}{r_{i+1}} \right)^{\frac{Q}{p}} \le C r^\lambda \tag{2.59}$$

thanks to (2.58). This means that $(u_j)_j$ is a Cauchy sequence and since u_j tends to $u(x)$ for almost every x thanks to Lebesgue's differentiation Theorem, letting j tend to infinity we get

$$|u(x) - u_i| \le C r^\lambda \quad \text{for every } i \in \mathbb{N}$$

which together with (2.51) and (2.53) concludes the proof. $\square$

Remark 2.11 The Korànyi balls satisfy the A-property.

References

1. Bonfiglioli, A., Lanconelli, E., Uguzzoni, F.: Stratified Lie Groups and Potential Theory for Their Sub-Laplacians. Springer Monographs in Mathematics. Springer, Berlin (2007)
2. Domokos, A.: Differentiability of solutions for the non-degenerate p-Laplacian in the Heisenberg group. J. Differ. Equ. **204**(2), 439–470 (2004)
3. Franchi, B.: BV spaces and rectifiability for Carnot-Carathéodory metrics: an introduction. NAFSA 7–Nonlinear Analysis. Function Spaces and Applications, vol. 7, pp. 72–132. Czech Academy of Sciences, Prague (2003)
4. Lu, G.: Embedding theorems into Lipschitz and BMO spaces and applications to quasilinear subelliptic differential equations. Publ. Mat. **40**(2), 301–329 (1996)
5. Varadarajan, V.S.: Lie Groups, Lie Algebras, and their Representations, vol. 102. Graduate Texts in Mathematics. Springer, New York (1984) (Reprint of the 1974 edition)

Chapter 3
The *p*-Laplace Equation

Abstract We give basic definitions and properties of the *p*-Laplace equation in the Heisenberg group. We establish existence and uniqueness results for the associated Dirichlet problem via variational methods and present some useful estimates. At the end we present the Hilbert–Haar existence theory for the variational functional associated to the *p*-Laplace equation which allows to prove that solutions to the non degenerate equation are Lipschitz continuous in domains satisfying a strict convexity condition.

Keywords *p*-Laplace equation · Weak solution · Bounded Slope Condition · Hilbert–Haar theory

3.1 Definitions and Notations

The *p*-Laplace equation, $1 < p < \infty$ is

$$\operatorname{div}_{\mathbb{H}}\left(\left(\delta^2 + |\nabla_{\mathbb{H}}u|^2\right)^{\frac{p-2}{2}} \nabla_{\mathbb{H}}u\right) = 0 \quad \text{in } \Omega. \tag{3.1}$$

It is the Euler–Lagrange equation for the *p*-Dirichlet functional

$$\mathcal{D}_p(u) = \frac{1}{p}\int_{\Omega}\left(\delta^2 + |\nabla_{\mathbb{H}}u|^2\right)^{\frac{p}{2}} \mathrm{d}x, \tag{3.2}$$

as will be explained in detail in Sect. 3.2.

We will use the term *non degenerate* for the case $\delta > 0$ and *degenerate* for $\delta = 0$.

We say that a function $u \in HW^{1,p}(\Omega)$ is a weak solution of (3.1) if the following holds

$$\int_{\Omega}\left(\delta^2 + |\nabla_{\mathbb{H}}u|^2\right)^{\frac{p-2}{2}} \langle\nabla_{\mathbb{H}}u, \nabla_{\mathbb{H}}\varphi\rangle\mathrm{d}x = 0 \quad \text{for all } \varphi \in C_0^{\infty}(\Omega). \tag{3.3}$$

Remark 3.1 By a density argument it can be seen that (3.3) holds also for all $\varphi \in HW_0^{1,p}(\Omega)$.

© The Author(s) 2015
D. Ricciotti, *p-Laplace Equation in the Heisenberg Group*,
SpringerBriefs in Mathematics, DOI 10.1007/978-3-319-23790-9_3

To simplify the notation we will write $z = (z_1, z_2) \in \mathbb{R}^2$ and

$$a_i(z) = \left(\delta^2 + |z|^2\right)^{\frac{p-2}{2}} z_i. \tag{3.4}$$

Next we will establish some bounds on the coefficients of the equation. Observe that

$$\begin{aligned}
\partial_{z_j} \left(\delta^2 + |z|^2\right)^{\frac{p-2}{2}} z_i &= (p-2)\left(\delta^2 + |z|^2\right)^{\frac{p-4}{2}} z_j z_i + \left(\delta^2 + |z|^2\right)^{\frac{p-2}{2}} \delta_{i,j} \\
&= \left(\delta^2 + |z|^2\right)^{\frac{p-4}{2}} \left((p-2)z_i z_j + \left(\delta^2 + |z|^2\right)\delta_{i,j}\right).
\end{aligned} \tag{3.5}$$

The eigenvalues of the matrix

$$M = \left(M_{i,j}\right)_{i,j=1,2} = \left((p-2)z_i z_j + \left(\delta^2 + |z|^2\right)\delta_{i,j}\right)_{i,j=1,2}$$

are

$$\begin{aligned}
\lambda_1(z) &= \left(\delta^2 + |z|^2\right), \\
\lambda_2(z) &= \delta^2 + |z|^2(p-1).
\end{aligned} \tag{3.6}$$

If $p \geq 2$ then the minimum eigenvalue is λ_1 and $\lambda_2 \leq (p-1)\left(\delta^2 + |z|^2\right)$.
   If $1 < p < 2$ then the minimun eigenvalue is λ_2 and $\lambda_2 \geq (p-1)\left(\delta^2 + |z|^2\right)$, therefore we have the following estimates

$$\left(\delta^2 + |z|^2\right)^{\frac{p-2}{2}} |\xi|^2 \leq \sum_{i,j=1}^{2} \partial_{z_j} a_i(z)\xi_i \xi_j \leq (p-1)\left(\delta^2 + |z|^2\right)^{\frac{p-2}{2}} |\xi|^2 \quad \text{if } p \geq 2$$

and

$$(p-1)\left(\delta^2 + |z|^2\right)^{\frac{p-2}{2}} |\xi|^2 \leq \sum_{i,j=1}^{2} \partial_{z_j} a_i(z)\xi_i \xi_j \leq \left(\delta^2 + |z|^2\right)^{\frac{p-2}{2}} |\xi|^2 \quad \text{if } 1 < p < 2.$$

Setting

$$c_p = \begin{cases} p-1 & \text{if } 1 < p < 2 \\ 1 & \text{if } p \geq 2 \end{cases} \tag{3.7}$$

and

$$C_p = \begin{cases} 1 & \text{if } 1 < p < 2 \\ p-1 & \text{if } p \geq 2 \end{cases} \tag{3.8}$$

we have the following estimates valid for $p > 1$:

$$|a_i(\nabla_{\mathbb{H}} u)| \leq \left(\delta^2 + |\nabla_{\mathbb{H}} u|^2\right)^{\frac{p-1}{2}},$$ (3.9)

$$\sum_{i,j=1}^{2} \partial_{z_j} a_i(\nabla_{\mathbb{H}} u)\xi_i \xi_j \geq c_p \left(\delta^2 + |\nabla_{\mathbb{H}} u|^2\right)^{\frac{p-2}{2}} |\xi|^2,$$ (3.10)

$$|\partial_{z_j} a_i(\nabla_{\mathbb{H}} u)| \leq C_p \left(\delta^2 + |\nabla_{\mathbb{H}} u|^2\right)^{\frac{p-2}{2}}.$$ (3.11)

Next we will provide some estimates for vectors in $\mathbb{R}^n$ that will be useful when dealing with the p-Laplace equation.

Lemma 3.1 *Let a and b be vectors in $\mathbb{R}^n$ and $p \geq 2$. Then*

$$\left\langle \left(\delta^2 + |b|^2\right)^{\frac{p-2}{2}} b - \left(\delta^2 + |a|^2\right)^{\frac{p-2}{2}} a, b - a \right\rangle \geq C_p |b - a|^p$$ (3.12)

where C_p is a positive constant depending only on p.

Proof A computation shows that

$$\frac{1}{2}\left(\left(\delta^2 + |b|^2\right)^{\frac{p-2}{2}} + \left(\delta^2 + |a|^2\right)^{\frac{p-2}{2}}\right)|b - a|^2$$

$$+ \frac{1}{2}\left(\left(\delta^2 + |b|^2\right)^{\frac{p-2}{2}} - \left(\delta^2 + |a|^2\right)^{\frac{p-2}{2}}\right)\left(|b|^2 - |a|^2\right)$$

$$= \frac{1}{2}\left(\left(\delta^2 + |b|^2\right)^{\frac{p-2}{2}} + \left(\delta^2 + |a|^2\right)^{\frac{p-2}{2}}\right)\langle b - a, b - a \rangle$$

$$+ \frac{1}{2}\left(\left(\delta^2 + |b|^2\right)^{\frac{p-2}{2}} - \left(\delta^2 + |a|^2\right)^{\frac{p-2}{2}}\right)\langle b + a, b - a \rangle$$

$$= \left\langle \left(\delta^2 + |b|^2\right)^{\frac{p-2}{2}} b - \left(\delta^2 + |a|^2\right)^{\frac{p-2}{2}} a, b - a \right\rangle.$$

Now observing that $\left(\left(\delta^2 + |b|^2\right)^{\frac{p-2}{2}} - \left(\delta^2 + |a|^2\right)^{\frac{p-2}{2}}\right)\left(|b|^2 - |a|^2\right)$ is always positive we get

$$\langle \left(\delta^2 + |b|^2\right)^{\frac{p-2}{2}} b - \left(\delta^2 + |a|^2\right)^{\frac{p-2}{2}} a, b - a \rangle \geq \frac{1}{2}\left(|b|^{p-2} + |a|^{p-2}\right)|b - a|^2 \geq 2^{2-p}|b - a|^p$$

since $p \geq 2$. □

The following lemma is taken from [1].

Lemma 3.2 *Let $a, b \in \mathbb{R}^n$ and $1 < p < \infty$. Then*

$$c_p \left(\delta^2 + |a|^2 + |b|^2\right)^{\frac{p-2}{2}} \leq \int_0^1 \left(\delta^2 + |ta + (1-t)b|^2\right)^{\frac{p-2}{2}} dt \leq C_p \left(\delta^2 + |a|^2 + |b|^2\right)^{\frac{p-2}{2}}$$

where c_p and C_p are two positive constants depending only on p.

Proof Suppose first that $|b| < |a|$. Then by the triangle inequality

$$|ta + (1-t)b| \geq t|a| - (1-t)|b|$$
$$\geq \frac{2}{3}|a| - \frac{1}{3}|b| \geq \frac{1}{3}|a| \tag{3.13}$$

if $\frac{2}{3} < t < 1$. Therefore we have

$$\delta^2 + |ta + (1-t)b|^2 \geq \delta^2 + \frac{1}{9}|a|^2$$
$$= \frac{1}{18}(18\delta^2 + 2|a|^2) \geq \frac{1}{18}\left(\delta^2 + |a|^2 + |b|^2\right). \tag{3.14}$$

The case $|a| \leq |b|$ is analogous.

Now

$$\delta^2 + |ta + (1-t)b|^2 \leq \delta^2 + 2t^2|a|^2 + 2(1-t)^2|b|^2 \leq 2\left(\delta^2 + |a|^2 + |b|^2\right)$$

so we have

$$\frac{1}{18^{\frac{p-2}{2}}}\left(\delta^2 + |a|^2 + |b|^2\right)^{\frac{p-2}{2}} \leq \left(\delta^2 + |ta + (1-t)b|^2\right)^{\frac{p-2}{2}} \leq 2^{\frac{p-2}{2}}\left(\delta^2 + |a|^2 + |b|^2\right)^{\frac{p-2}{2}}$$

if $p \geq 2$ and

$$\frac{1}{2^{\frac{p-2}{2}}}\left(\delta^2 + |a|^2 + |b|^2\right)^{\frac{p-2}{2}} \leq \left(\delta^2 + |ta + (1-t)b|^2\right)^{\frac{p-2}{2}} \leq 18^{\frac{p-2}{2}}\left(\delta^2 + |a|^2 + |b|^2\right)^{\frac{p-2}{2}}$$

if $1 < p < 2$.

Plugging these into the integral gives the desired result.   □

Lemma 3.3 *Let $a, b \in \mathbb{R}^n$ and $1 < p < \infty$. Then*

$$c_p\left(\delta^2 + |a|^2 + |b|^2\right)^{\frac{p-2}{2}}|b - a|^2 \leq \langle\left(\delta^2 + |a|^2\right)^{\frac{p-2}{2}}a - \left(\delta^2 + |b|^2\right)^{\frac{p-2}{2}}b, a - b\rangle$$
$$\leq C_p\left(\delta^2 + |a|^2 + |b|^2\right)^{\frac{p-2}{2}}|b - a|^2$$

where c_p and C_p are two positive constants depending only on p.

Proof Let $f(z) = (\delta^2 + |z|^2)^{\frac{p}{2}}$. Then the Euclidean gradient is $\nabla f(z) = \nabla_z f(z) = p(\delta^2 + |z|^2)^{\frac{p-2}{2}}z$. Using vector notation we have

$$\langle \nabla f(a) - \nabla f(b), a - b \rangle = \int_0^1 \left\langle \frac{d}{dt} \nabla f(ta + (1 - t)b), a - b \right\rangle dt$$

$$= \int_0^1 \langle \nabla^2 f(ta + (1 - t)b)(a - b), a - b \rangle \, dt \quad (3.15)$$

and by (3.10)

$$c_p \int_0^1 \left(\delta^2 + |ta + (1 - t)b|^2 \right)^{\frac{p-2}{2}} |a - b|^2 dt \le \langle \nabla f(a) - \nabla f(b), a - b \rangle$$

$$\le C_p \int_0^1 \left(\delta^2 + |ta + (1 - t)b|^2 \right)^{\frac{p-2}{2}} |a - b|^2 dt \quad (3.16)$$

and now we conclude by Lemma 3.2. □

Lemma 3.4 *Let $a, b \in \mathbb{R}^n$ and $1 < p < \infty$. Then*

$$\left| \left(\delta^2 + |b|^2 \right)^{\frac{p-2}{2}} b - \left(\delta^2 + |a|^2 \right)^{\frac{p-2}{2}} a \right| \le C_p |b - a| \left(\delta^2 + |a|^2 + |b|^2 \right)^{\frac{p-2}{2}} \quad (3.17)$$

where C_p is a constant depending only on p.

Proof By the fundamental theorem of calculus we have

$$\left| \left(\delta^2 + |b|^2 \right)^{\frac{p-2}{2}} b - \left(\delta^2 + |a|^2 \right)^{\frac{p-2}{2}} a \right|$$

$$= \left| \int_0^1 \frac{d}{dt} \left(\delta^2 + |tb + (1 - t)a|^2 \right)^{\frac{p-2}{2}} (tb + (1 - t)a) \, dt \right|$$

$$\le (p - 2) \int_0^1 \left(\delta^2 + |tb + (1 - t)a|^2 \right)^{\frac{p-2}{2} - 1} |tb + (1 - t)a| |b - a| \, dt$$

$$+ \int_0^1 \left(\delta^2 + |tb + (1 - t)a|^2 \right)^{\frac{p-2}{2}} |b - a| \, dt$$

$$\le (p - 1)|b - a| \int_0^1 \left(\delta^2 + |tb + (1 - t)a|^2 \right)^{\frac{p-2}{2}} dt \quad (3.18)$$

and now we can conclude from Lemma 3.2. □

Lemma 3.5 *Let $a, b \in \mathbb{R}^n$ and $1 < p < 2$. Then*

$$\left| \left(\delta^2 + |b|^2 \right)^{\frac{p-2}{2}} b - \left(\delta^2 + |a|^2 \right)^{\frac{p-2}{2}} a \right| \le C_p |b - a|^{\frac{p-1}{2}} \quad (3.19)$$

where C_p is a positive constant depending only on p.

Lemma 3.6 *Let $u_k \in HW^{1,p}(\Omega)$ be a sequence of weak solutions of Eq.(3.1) with fixed $\delta \geq 0$, $1 < p < \infty$. If $u_k \xrightarrow{w} u$ in $HW^{1,p}(\Omega)$ then there exists a subsequence u_{k_j} such that $\nabla_{\mathbb{H}} u_{k_j} \to \nabla_{\mathbb{H}} u$ in $L^p_{loc}(\Omega)$ and u is a solution of the equation.*

Proof Since the u_ks are solutions they satisfy

$$\int_{\Omega} \left(\delta^2 + |\nabla_{\mathbb{H}} u_k|^2\right)^{\frac{p-2}{2}} \langle \nabla_{\mathbb{H}} u_k, \nabla_{\mathbb{H}} \varphi \rangle \, dx = 0 \quad \text{for all } \varphi \in HW^{1,p}_0(\Omega). \quad (3.20)$$

Adding and subtracting the same term we get

$$\int_{\Omega} \langle (\delta^2 + |\nabla_{\mathbb{H}} u|^2)^{\frac{p-2}{2}} \nabla_{\mathbb{H}} u - (\delta^2 + |\nabla_{\mathbb{H}} u_k|^2)^{\frac{p-2}{2}} \nabla_{\mathbb{H}} u_k, \nabla_{\mathbb{H}} \varphi \rangle \, dx$$
$$= \int_{\Omega} \left(\delta^2 + |\nabla_{\mathbb{H}} u|^2\right)^{\frac{p-2}{2}} \langle \nabla_{\mathbb{H}} u, \nabla_{\mathbb{H}} \varphi \rangle \, dx. \quad (3.21)$$

Now using $\varphi = \xi(u_k - u)$ as a test function in (3.21) (where $\xi \in C_0^\infty(\Omega)$) the right hand side becomes

$$\int_{\Omega} \left(\delta^2 + |\nabla_{\mathbb{H}} u|^2\right)^{\frac{p-2}{2}} \langle \nabla_{\mathbb{H}} u, \nabla_{\mathbb{H}}(\xi(u_k - u)) \rangle \, dx$$
$$= \int_{\Omega} (u_k - u) \left(\delta^2 + |\nabla_{\mathbb{H}} u|^2\right)^{\frac{p-2}{2}} \langle \nabla_{\mathbb{H}} u, \nabla_{\mathbb{H}} \xi \rangle \, dx$$
$$+ \int_{\Omega} \xi \left(\delta^2 + |\nabla_{\mathbb{H}} u|^2\right)^{\frac{p-2}{2}} \langle \nabla_{\mathbb{H}} u, \nabla_{\mathbb{H}} u_k - \nabla_{\mathbb{H}} u \rangle \, dx \quad (3.22)$$

and since

$$\left(\delta^2 + |\nabla_{\mathbb{H}} u|^2\right)^{\frac{p-2}{2}} \langle \nabla_{\mathbb{H}} u, \nabla_{\mathbb{H}} \xi \rangle \in L^{\frac{p}{p-1}}(\Omega)$$

and

$$\xi \left(\delta^2 + |\nabla_{\mathbb{H}} u|^2\right)^{\frac{p-2}{2}} \nabla_{\mathbb{H}} u \in L^{\frac{p}{p-1}}(\Omega, \mathbb{R}^n)$$

by weak convergence the right hand side of (3.21) tends to zero. So we get

$$\lim_{k \to \infty} \int_{\Omega} \langle \left(\delta^2 + |\nabla_{\mathbb{H}} u|^2\right)^{\frac{p-2}{2}} \nabla_{\mathbb{H}} u - \left(\delta^2 + |\nabla_{\mathbb{H}} u_k|^2\right)^{\frac{p-2}{2}} \nabla_{\mathbb{H}} u_k, \nabla_{\mathbb{H}}(\xi(u_k - u)) \rangle \, dx = 0. \quad (3.23)$$

Denoting by I_k the integrals on the left hand side of (3.23) we have

$$I_k = \int_\Omega \xi \langle (\delta^2 + |\nabla_\mathbb{H} u|^2)^{\frac{p-2}{2}} \nabla_\mathbb{H} u - (\delta^2 + |\nabla_\mathbb{H} u_k|^2)^{\frac{p-2}{2}} \nabla_\mathbb{H} u_k, \nabla_\mathbb{H} u_k - \nabla_\mathbb{H} u \rangle \, dx$$

$$+ \int_\Omega (u_k - u) \langle (\delta^2 + |\nabla_\mathbb{H} u|^2)^{\frac{p-2}{2}} \nabla_\mathbb{H} u - (\delta^2 + |\nabla_\mathbb{H} u_k|^2)^{\frac{p-2}{2}} \nabla_\mathbb{H} u_k, \nabla_\mathbb{H} \xi \rangle \, dx$$

$$= I_{k,1} + I_{k,2}. \tag{3.24}$$

Now by Hölder's inequality we have

$$I_{k,2} \le \left(\int_\Omega |\nabla_\mathbb{H} \xi|^p |u_k - u|^p dx \right)^{\frac{1}{p}} \int_{\mathrm{supp}\xi} \left| (\delta^2 + |\nabla_\mathbb{H} u|^2)^{\frac{p-1}{2}} + (\delta^2 + |\nabla_\mathbb{H} u_k|^2)^{\frac{p-1}{2}} \right|^{\frac{p}{p-1}} dx$$

$$\le C_p \left(\int_\Omega |\nabla_\mathbb{H} \xi|^p |u_k - u|^p dx \right)^{\frac{1}{p}} \left(\int_{\mathrm{supp}\xi} (\delta^2 + |\nabla_\mathbb{H} u|^2)^{\frac{p}{2}} dx + \int_{\mathrm{supp}\xi} (\delta^2 + |\nabla_\mathbb{H} u_k|^2)^{\frac{p}{2}} dx \right).$$

Since u_k is weakly convergent in $HW^{1,p}(\Omega)$ the last integral is uniformly bounded by a constant independent from k; moreover by the compact embedding of Theorem 2.9 there exists a subsequence such that

$$u_{k_j} \nabla_\mathbb{H} \xi \to u \nabla_\mathbb{H} \xi \quad \text{in } L^p(\Omega)$$

which implies that $I_{k_j,2} \to 0$. Together with (3.23) and (3.24) this gives

$$\lim_{j \to \infty} I_{k_j,1} = 0. \tag{3.25}$$

If $p \ge 2$ using Lemma 3.1 we get

$$I_{k_j,1} \ge \int_{\mathrm{supp}\xi} |\nabla_\mathbb{H} u_{k_j} - \nabla_\mathbb{H} u|^p dx$$

so by (3.25) $\nabla_\mathbb{H} u_{k_j} \to \nabla_\mathbb{H} u$ in $L^p_{loc}(\Omega)$ for $p \ge 2$.

Now that we have strong convergence we can prove that u is indeed a solution of Eq. (3.1). Using Lemma 3.4 and Hölder's inequality we get

$$\int_\Omega \langle (\delta^2 + |\nabla_\mathbb{H} u|^2)^{\frac{p-2}{2}} \nabla_\mathbb{H} u - (\delta^2 + |\nabla_\mathbb{H} u_k|^2)^{\frac{p-2}{2}} \nabla_\mathbb{H} u_{k_j}, \nabla_\mathbb{H} \varphi \rangle \, dx$$

$$\le (p-1) \|\nabla_\mathbb{H} \varphi\|_{L^\infty(\Omega)} \int_\Omega |\nabla_\mathbb{H} u - \nabla_\mathbb{H} u_k| (\delta^2 + |\nabla_\mathbb{H} u|^2 + |\nabla_\mathbb{H} u_k|^2)^{\frac{p-2}{2}} \, dx$$

$$\le C_{p,\varphi} \|\nabla_\mathbb{H} u - \nabla_\mathbb{H} u_j\|_{L^p(\mathrm{supp}\varphi)} \left(\int_\Omega |\nabla_\mathbb{H} u|^{\frac{p(p-2)}{p-1}} + |\nabla_\mathbb{H} u_k|^{\frac{p(p-2)}{p-1}} dx \right)^{\frac{p-1}{p}}$$

which tends to zero since the last integral is uniformly bounded with respect to k.

If $1 < p < 2$ by Lemma 3.3 we have

$$I_{k_j,1} \geq \int_\Omega \xi \left(\delta^2 + |\nabla_{\mathbb{H}} u|^2 + |\nabla_{\mathbb{H}} u_{k_j}|^2\right)^{\frac{p-2}{2}} |\nabla_{\mathbb{H}} u_{k_j} - \nabla_{\mathbb{H}} u|^2 dx.$$

Now use

$$\int_\Omega f^p dx = \int_\Omega f^p W^{\frac{p(p-2)}{2}} W^{\frac{p(2-p)}{2}} dx \leq \left(\int_\Omega f^2 W^{p-2} dx\right)^{\frac{p}{2}} \left(\int_\Omega W^p dx\right)^{1-\frac{p}{2}}$$

for $f = |\nabla_{\mathbb{H}} u_k - \nabla_{\mathbb{H}} u|$ and $W = \left(\delta^2 + |\nabla_{\mathbb{H}} u|^2 + |\nabla_{\mathbb{H}} u_{k_j}|^2\right)^{\frac{1}{2}}$. Since

$$\int_\Omega \left(\delta^2 + |\nabla_{\mathbb{H}} u|^2 + |\nabla_{\mathbb{H}} u_{k_j}|^2\right)^{\frac{p}{2}} dx$$

is bounded, by (3.25) we get the strong convergence also in the case $1 < p < 2$. To prove that u is a solution it will be sufficient to use Lemma 3.5 and the L^p convergence just proved. $\square$

Definition 3.1 Let $u \in HW^{1,p}(\Omega)$. We say that $u \leq 0$ in $\partial\Omega$ in the Sobolev sense if $u^+ = \max\{u, 0\} \in HW_0^{1,p}(\Omega)$. We say that $u \geq 0$ in $\partial\Omega$ in the Sobolev sense if $-u \leq 0$ in $\partial\Omega$ in the Sobolev sense. Finally if $u \in HW^{1,p}(\Omega)$, we say that $u \leq v$ in $\partial\Omega$ in the Sobolev sense if $u - v \leq 0$ in $\partial\Omega$ in the Sobolev sense.

Theorem 3.1 (Weak Comparison principle) *Let $u, v \in HW^{1,p}(\Omega)$ be weak solutions of Eq. (3.1) for $\delta \geq 0$ and $1 < p < \infty$. If $u \geq v$ in $\partial\Omega$ in the Sobolev sense then $u \geq v$ a.e in Ω.*

Proof Since u and v are solutions they satisfy

$$\int_\Omega \left(\delta^2 + |\nabla_{\mathbb{H}} u|^2\right)^{\frac{p-2}{2}} \langle \nabla_{\mathbb{H}} u, \nabla_{\mathbb{H}} \varphi \rangle dx = 0 \quad \text{for all } \varphi \in HW_0^{1,p}(\Omega) \qquad (3.26)$$

$$\int_\Omega \left(\delta^2 + |\nabla_{\mathbb{H}} v|^2\right)^{\frac{p-2}{2}} \langle \nabla_{\mathbb{H}} v, \nabla_{\mathbb{H}} \varphi \rangle dx = 0 \quad \text{for all } \varphi \in HW_0^{1,p}(\Omega). \qquad (3.27)$$

Subtracting the previous equalities and choosing

$$\varphi = (v - u)^+ = \max\{v - u, 0\}$$

which is an admissible function because $u \geq v$ in $\partial\Omega$ in the Sobolev sense, for $p \geq 2$ in virtue of Lemma 3.1 we get

$$0 = \int_{\{v > u\}} \langle \left(\delta^2 + |\nabla_{\mathbb{H}} v|^2\right)^{\frac{p-2}{2}} \nabla_{\mathbb{H}} v - \left(\delta^2 + |\nabla_{\mathbb{H}} u|^2\right)^{\frac{p-2}{2}} \nabla_{\mathbb{H}} u, \nabla_{\mathbb{H}} v - \nabla_{\mathbb{H}} u \rangle dx$$

$$\geq C_p \int_{\{v > u\}} |\nabla_{\mathbb{H}} v - \nabla_{\mathbb{H}} u|^p dx \qquad (3.28)$$

where we have denoted $\{v > u\} = \{x \in \Omega \mid v(x) > u(x)\}$. This implies

$$\int_{\{v>u\}} |\nabla_{\mathbb{H}} v - \nabla_{\mathbb{H}} u|^p dx = 0$$

which means that either $|\{v > u\}| = 0$ or $\nabla_{\mathbb{H}} u = \nabla_{\mathbb{H}} v$ a.e in $\{v > u\}$. This last case would imply that $v - u$ is constant on $\{v > u\}$ and keeping in mind that $(v - u)^+ \in HW_0^{1,p}(\Omega)$ this constant should be zero, which is a contradiction. So we must have $|\{v > u\}| = 0$ which is precisely the result stated in the lemma for the case $p \geq 2$.

Consider now $1 < p < 2$. If $\nabla_{\mathbb{H}} u = \nabla_{\mathbb{H}} v = 0$ in $\{v > u\}$ then arguing as above we get a contradiction. This means that

$$\{v > u\} = \{v > u\} \cap \{\nabla_{\mathbb{H}} u = \nabla_{\mathbb{H}} v = 0\}^c = A .$$

Now using Lemma 3.3 we get

$$
\begin{aligned}
0 &= \int_{\{v>u\}} \left\langle \left(\delta^2 + |\nabla_{\mathbb{H}} v|^2\right)^{\frac{p-2}{2}} \nabla_{\mathbb{H}} v - \left(\delta^2 + |\nabla_{\mathbb{H}} u|^2\right)^{\frac{p-2}{2}} \nabla_{\mathbb{H}} u, \nabla_{\mathbb{H}} v - \nabla_{\mathbb{H}} u \right\rangle dx \\
&\geq \int_A \left(\delta^2 + |\nabla_{\mathbb{H}} u|^2 + |\nabla_{\mathbb{H}} v|^2\right)^{\frac{p-2}{2}} |\nabla_{\mathbb{H}} v - \nabla_{\mathbb{H}} u|^2 dx \qquad (3.29)
\end{aligned}
$$

which implies that either $\left(\delta^2 + |\nabla_{\mathbb{H}} u|^2 + |\nabla_{\mathbb{H}} v|^2\right)^{\frac{p-2}{2}} |\nabla_{\mathbb{H}} v - \nabla_{\mathbb{H}} u|^2 = 0$ a.e in A or $|A| = 0$. Since in this set at least one of the norms of the gradients is stricly positive we get that $|\nabla_{\mathbb{H}} v - \nabla_{\mathbb{H}} u|^2 = 0$ a.e in $\{v > u\}$, which is again a contradiction. So $|A| = 0$ which completely proves the statement of the theorem. $\qquad \square$

3.2 Existence and Uniqueness

The p-Laplace equation is the Euler–Lagrange equation for the p-Dirichlet functional

$$\mathcal{D}_{p,\delta}(u) = \frac{1}{p} \int_\Omega \left(\delta^2 + |\nabla_{\mathbb{H}} u|^2\right)^{\frac{p}{2}} dx. \qquad (3.30)$$

More precisely we have

Theorem 3.2 *Let* $\psi \in HW^{1,p}(\Omega)$ *and* $\mathcal{A}_\psi = \left\{ v \in HW^{1,p}(\Omega) \mid v - \psi \in HW_0^{1,p}(\Omega) \right\}$. *Then u is a weak solution of the Dirichlet problem*

$$
\begin{cases}
div_{\mathbb{H}} \left(\left(\delta^2 + |\nabla_{\mathbb{H}} u|^2\right)^{\frac{p-2}{2}} \nabla_{\mathbb{H}} u \right) = 0 & in \ \Omega \\
u - \psi \in HW_0^{1,p}(\Omega)
\end{cases}
\qquad (3.31)
$$

if and only if u is a minimum of the p-Dirichlet functional $\mathcal{D}_{p,\delta}$ *in* $\mathcal{A}_\psi$.

Proof Let u be a minimum of the p-Dirichlet functional $\mathcal{D}_{p,\delta}$. This means that $\mathcal{D}_{p,\delta}(u) \leq \mathcal{D}_{p,\delta}(u + t\varphi)$ for all $\varphi \in C_0^\infty(\Omega)$ and $t \in \mathbb{R}$ (note that $u + t\varphi \in \mathcal{A}_\psi$). Therefore the real variable function $F(t) = \mathcal{D}_{p,\delta}(u + t\varphi)$ has a minimum at $t = 0$.

Now

$$
\begin{aligned}
\frac{F(t) - F(0)}{t} &= \frac{1}{p} \int_\Omega \frac{(\delta^2 + |\nabla_{\mathbb{H}} u + t\nabla_{\mathbb{H}}\varphi|^2)^{\frac{p}{2}} - (\delta^2 + |\nabla_{\mathbb{H}} u|^2)^{\frac{p}{2}}}{t} \, dx \\
&= \frac{1}{p} \int_\Omega \frac{f(\nabla_{\mathbb{H}} u + t\nabla_{\mathbb{H}}\varphi) - f(\nabla_{\mathbb{H}} u)}{t} \, dx
\end{aligned}
\tag{3.32}
$$

where we have denoted $f(z) = (\delta^2 + |z|^2)^{\frac{p}{2}}$, $z \in \mathbb{R}^2$.

The integrand of (3.32) converges pointwisely to

$$
\langle \nabla f(\nabla_{\mathbb{H}} u), \nabla_{\mathbb{H}}\varphi \rangle = p(\delta^2 + |\nabla_{\mathbb{H}} u|^2)^{\frac{p-2}{2}} \langle \nabla_{\mathbb{H}} u, \nabla_{\mathbb{H}}\varphi \rangle
$$

(where $\nabla f(z) = \nabla_z f(z) = p(\delta^2 + |z|^2)^{\frac{p-2}{2}} z$ is the Euclidean gradient). In order to pass to the limit under the integral we estimate the integrand of (3.32) considering $t < 1$. Therefore

$$
\begin{aligned}
\frac{1}{t} &|f(\nabla_{\mathbb{H}} u + t\nabla_{\mathbb{H}}\varphi) - f(\nabla_{\mathbb{H}} u)| \\
&= \frac{1}{t} \left| \langle \nabla f(\nabla_{\mathbb{H}} u + \bar{t}\nabla_{\mathbb{H}}\varphi), t\nabla_{\mathbb{H}}\varphi \rangle \right| \\
&\leq \|\nabla_{\mathbb{H}}\varphi\|_{L^\infty(\Omega)} \left| \nabla f(\nabla_{\mathbb{H}} u + \bar{t}\nabla_{\mathbb{H}}\varphi) \right| \\
&\leq p \|\nabla_{\mathbb{H}}\varphi\|_{L^\infty(\Omega)} (\delta^2 + |\nabla_{\mathbb{H}} u + \bar{t}\nabla_{\mathbb{H}}\varphi|^2)^{\frac{p-2}{2}} |\nabla_{\mathbb{H}} u + \bar{t}\nabla_{\mathbb{H}}\varphi| \\
&\leq p \|\nabla_{\mathbb{H}}\varphi\|_{L^\infty(\Omega)} (\delta^2 + |\nabla_{\mathbb{H}} u + \bar{t}\nabla_{\mathbb{H}}\varphi|^2)^{\frac{p-1}{2}} \\
&\leq C_{p,\varphi}(\delta^{p-1} + |\nabla_{\mathbb{H}} u|^{p-1}) \in L^1(\Omega)
\end{aligned}
\tag{3.33}
$$

where we have used the mean value theorem with $0 < \bar{t} < t < 1$. By Lebesgue's dominated convergence theorem we obtain

$$
F'(0) = \int_\Omega (\delta^2 + |\nabla_{\mathbb{H}} u|^2)^{\frac{p-2}{2}} \langle \nabla_{\mathbb{H}} u, \nabla_{\mathbb{H}}\varphi \rangle dx = 0
\tag{3.34}
$$

because $t = 0$ is a minimum for the function F. Equality (3.34) holds for all $\varphi \in C_0^\infty(\Omega)$ proving the first part of the theorem.

Conversely let u be a weak solution of (3.31) and $v \in \mathcal{A}_\psi$. Since f is a convex function we have with the previuos notation:

$$
f(\nabla_{\mathbb{H}} v) \geq f(\nabla_{\mathbb{H}} u) + \langle \nabla f(\nabla_{\mathbb{H}} u), \nabla_{\mathbb{H}} u - \nabla_{\mathbb{H}} v \rangle.
\tag{3.35}
$$

This implies

$$\frac{1}{p} \int_\Omega \left(\delta^2 + |\nabla_\mathbb{H} v|^2 \right)^{\frac{p}{2}} dx \geq \frac{1}{p} \int_\Omega \left(\delta^2 + |\nabla_\mathbb{H} u|^2 \right)^{\frac{p}{2}} dx$$

$$+ \frac{1}{p} \int_\Omega \left(\delta^2 + |\nabla_\mathbb{H} u|^2 \right)^{\frac{p}{2}} \langle \nabla_\mathbb{H} u, \nabla_\mathbb{H} (v - u) \rangle dx \qquad (3.36)$$

and since $v - u \in HW_0^{1,p}(\Omega)$ the last integral is zero because u is a weak solution of (3.31), so $\mathcal{D}_{p,\delta}(u) \leq \mathcal{D}_{p,\delta}(v)$ for all $v \in \mathcal{A}_\psi$. $\qquad\qquad\square$

Using the direct method of the calculus of variations we get the existence and uniqueness for the solution of the Dirichlet problem for the p-Laplace equation by proving existence and uniqueness for the minimun of the p-Dirichlet functional.

Theorem 3.3 *Let $1 < p < \infty$, $\delta \geq 0$. There exists a unique solution of the Dirichlet problem (3.31).*

Proof We prove first uniqueness. Let u_1 and u_2 be solutions of (3.31). Then for all $\varphi \in HW_0^{1,p}(\Omega)$ we have

$$\int_\Omega \left(\delta^2 + |\nabla_\mathbb{H} u_1|^2 \right)^{\frac{p-2}{2}} \langle \nabla_\mathbb{H} u_1, \nabla_\mathbb{H} \varphi \rangle dx = 0 \qquad (3.37)$$

$$\int_\Omega \left(\delta^2 + |\nabla_\mathbb{H} u_2|^2 \right)^{\frac{p-2}{2}} \langle \nabla_\mathbb{H} u_2, \nabla_\mathbb{H} \varphi \rangle dx = 0. \qquad (3.38)$$

Subtracting the previous relations we get

$$\int_\Omega \langle \left(\delta^2 + |\nabla_\mathbb{H} u_1|^2 \right)^{\frac{p-2}{2}} \nabla_\mathbb{H} u_1 - \left(\delta^2 + |\nabla_\mathbb{H} u_2|^2 \right)^{\frac{p-2}{2}} \nabla_\mathbb{H} u_2, \nabla_\mathbb{H} \varphi \rangle dx = 0 \qquad (3.39)$$

valid for all $\varphi \in HW_0^{1,p}(\Omega)$. In particular we can choose $\varphi = u_1 - u_2$ to get

$$0 = \int_\Omega \langle \left(\delta^2 + |\nabla_\mathbb{H} u_1|^2 \right)^{\frac{p-2}{2}} \nabla_\mathbb{H} u_1 - \left(\delta^2 + |\nabla_\mathbb{H} u_2|^2 \right)^{\frac{p-2}{2}} \nabla_\mathbb{H} u_2, \nabla_\mathbb{H} u_1 - \nabla_\mathbb{H} u_2 \rangle dx$$

$$\geq c_p \int_\Omega |\nabla_\mathbb{H} u_1 - \nabla_\mathbb{H} u_2|^p dx \qquad (3.40)$$

if $p \geq 2$ by Lemma 3.1. We deduce that $\nabla_\mathbb{H} u_1 - \nabla_\mathbb{H} u_2 = 0$ a.e in Ω which implies that $u_1 - u_2 = c$ and the costant c must be zero because $u_1 - u_2 \in HW_0^{1,p}(\Omega)$. This proves uniqueness in the case $p \geq 2$.

If $1 < p < 2$ and $\nabla_\mathbb{H} u_1 = \nabla_\mathbb{H} u_2 = 0$ in Ω then arguing as above we get $u_1 = u_2$ a.e. in $\Omega \cap \{\nabla_\mathbb{H} u_1 = \nabla_\mathbb{H} u_2 = 0\}$. Denote $A = \Omega \cap \{\nabla_\mathbb{H} u_1 = \nabla_\mathbb{H} u_2 = 0\}^c$ and use Lemma 3.3 to get

$$0 = \int_\Omega \langle (\delta^2 + |\nabla_{\mathbb{H}} u_1|^2)^{\frac{p-2}{2}} \nabla_{\mathbb{H}} u_1 - (\delta^2 + |\nabla_{\mathbb{H}} u_2|^2)^{\frac{p-2}{2}} \nabla_{\mathbb{H}} u_2, \nabla_{\mathbb{H}} u_1 - \nabla_{\mathbb{H}} u_2 \rangle dx$$

$$\geq c_p \int_\Omega (\delta^2 + |\nabla_{\mathbb{H}} u_1|^2 + |\nabla_{\mathbb{H}} u_2|^2)^{\frac{p-2}{2}} |\nabla_{\mathbb{H}} u_1 - \nabla_{\mathbb{H}} u_2|^2 dx \qquad (3.41)$$

which implies that $(\delta^2 + |\nabla_{\mathbb{H}} u_1|^2 + |\nabla_{\mathbb{H}} u_2|^2)^{\frac{p-2}{2}} |\nabla_{\mathbb{H}} u_1 - \nabla_{\mathbb{H}} u_2|^2 = 0$ a.e. in A. Since in this set at least one of the norms of the gradients is strictly positive we get that $|\nabla_{\mathbb{H}} u_1 - \nabla_{\mathbb{H}} u_2|^2 = 0$ a.e in A, so $u_1 = u_2$ a.e in Ω.

Now let

$$\lambda = \inf_{u \in \mathcal{A}_\psi} \mathcal{D}_{p,\delta}$$

and observe that $0 \leq \lambda \leq \mathcal{D}_{p,\delta}(\psi) < \infty$.

Consider a minimizing sequence $u_n \in \mathcal{A}_\psi$ such that

$$\mathcal{D}_{p,\delta}(u_n) \leq \lambda + \frac{1}{n}. \qquad (3.42)$$

This implies that

$$\int_\Omega |\nabla_{\mathbb{H}} u_n|^p dx \leq p \mathcal{D}_{p,\delta}(u_n) \leq p\left(\lambda + \frac{1}{n}\right) \leq p(\lambda + 1) \quad \text{for all } n. \qquad (3.43)$$

Since Ω is bounded the L^p norm of the gradient of a function is equivalent to the $HW^{1,p}(\Omega)$ norm: this means that $\|u_n\|_{HW^{1,p}(\Omega)}$ is uniformly bounded with respect to the index n. Since $1 < p < \infty$ the space $HW^{1,p}(\Omega)$ is reflexive and we can extract a converging subsequence (which we denote again by u_n), namely there exists $v \in HW^{1,p}(\Omega)$ such that $u_n \xrightarrow{w} v$ in $HW^{1,p}(\Omega)$. Moreover $v \in \mathcal{A}_\psi$ because $HW_0^{1,p}(\Omega)$ is closed under weak convergence. Now by weakly lower semicontinuity of the functional $\mathcal{D}_{p,\delta}$ we get

$$\mathcal{D}_{p,\delta}(v) \leq \liminf_{n \to \infty} \mathcal{D}_{p,\delta}(u_n) \leq \lim_{n \to \infty}\left(\lambda + \frac{1}{n}\right) = \lambda$$

so $v \in \mathcal{A}_\psi$ is a minimum for $\mathcal{D}_{p,\delta}$. $\qquad \square$

3.3 Hilbert–Haar Existence Theory

The Hilbert–Haar existence theory allows to prove the existence of Lipschitz minimizers for certain types of convex functionals. It has already been used in the Euclidean setting and we will use it as the first step in proving regularity of solutions to the p-Laplace equation. We start defining what in the literature is known as Bounded Slope Condition.

Definition 3.2 *(B.S.C)* Let D be a bounded open set of $\mathbb{H}$ and $\psi : \partial D \to \mathbb{R}$. Then ψ satisfies the Bounded Slope Condition (of rank K) if for all $y \in \partial D$ there exist two affine functions (in the Euclidean sense) L_y^+ and L_y^- satisfying the following properties:

1. $L_y^-(y) = \psi(y) = L_y^+(y)$,
2. $L_y^-(x) \le \psi(x) \le L_y^+(x)$ for all $x \in \partial D$,
3. L_y^+ and L_y^- are Lipschitz functions relatively to the Carnot-Carathéodory metric with Lipschitz constant K.

This condition will be useful when ψ is the boundary value of the Dirichlet problem. Convexity of the boundary in the Euclidean sense is not a sufficient condition in general for having a B.S.C, we need to assume for example that the boundary is sufficiently curved. More precisely we have to ask that D is a bounded open set and that

$$\text{there exists } \Gamma > 0 \text{ s.t for all } y \in \partial D \text{ there exists } b_y \in \mathbb{S}^2 :$$
$$\langle x - y, b_y \rangle \ge \Gamma |x - y|_E^2 \tag{3.44}$$

where $\langle \cdot \rangle$ and $| \cdot |_E$ are the standard Euclidean scalar product and norm and we are denoting by x and y points in $\mathbb{R}^3$. The next result is due to Miranda [2].

Theorem 3.4 *Let D be an open bounded set satisfying (3.44). If $\psi \in C^2(\overline{D})$ then it satisfies the B.S.C.*

Proof Take

$$L_y^+(x) = L^+(x) = \psi(y) + \langle \nabla \psi(y), x - y \rangle + K \langle b, x - y \rangle \tag{3.45}$$
$$L_y^-(x) = L^+(x) = \psi(y) + \langle \nabla \psi(y), x - y \rangle - K \langle b, x - y \rangle \tag{3.46}$$

where $x \in D$, K is a constant to be chosen later and $b = b_y$ is the vector from (3.44).

Since $\psi \in C^2(\overline{D})$, the Taylor formula with Lagrange's remainder gives the existence of $t \in [0, 1]$ and $\xi = tx + (1 - t)y$ such that

$$\psi(x) = \psi(y) + \langle \nabla \psi(y), x - y \rangle + \frac{1}{2} \langle \nabla^2 \psi(\xi)(x - y), x - y \rangle$$
$$\le \psi(y) + \langle \nabla \psi(y), x - y \rangle + \frac{(2n + 1)^2}{2} \left\| \nabla^2 \psi \right\|_{L^\infty(\overline{D})} |x - y|^2. \tag{3.47}$$

Now choosing $K = \frac{(2n+1)^2}{2\Gamma} \left\| \nabla^2 \psi \right\|_{L^\infty(\overline{D})}$ in (3.45) and using property (3.44) we get that $\psi(x) \le L^+(x)$. Analogously we have that with the same choice of K it holds $L^-(x) \le \psi(x)$.

Observe that

$$|L^+(x) - L^+(z)| = |L^-(x) - L^-(z)| = |\langle \nabla \psi(y), x - z \rangle + K \langle b, x - z \rangle|$$
$$\le (|\nabla \psi(y)| + K)|x - z|_E \le (|\nabla \psi(y)| + K)(\operatorname{diam} D)^{\frac{1}{2}} |x - z|_K$$

where $|\cdot|$ is the standard Euclidean norm and $|\cdot|_K$ is the Kornyi norm which is equivalent to the Carnot-Carathéodory norm (Proposition 2.2), so L^+ and L^- satisfy all the requirements of Definition 3.2 and the theorem is proved. □

Remark 3.2 Note that the Lipschitz constant of L^+ and L^- depends on n, Γ, $\|\nabla\psi\|_{L^\infty}$, $\|\nabla^2\psi\|_{L^\infty}$ and diam D.

The following theorem is taken from [4].

Theorem 3.5 *Let D be a bounded open set satisfying (3.44) and $u \in HW^{1,p}(D)$ a weak solution of the Dirichlet problem (3.31) with boundary datum $\psi \in C^2(\overline{D})$. Then there exists $M = M_{n,\Gamma,|\nabla\psi|,|\nabla^2\psi|,\mathrm{diam}\,D} > 0$ such that*

$$\|\nabla_{\mathbb{H}} u\|_{L^\infty(D)} \le M.$$

Proof First we will prove that

$$\sup_{\substack{x\in\overline{D}\\y\in\partial D}} \frac{|u(x) - u(y)|}{d(x,y)} \le M.$$

By Theorem 3.4 ψ satisfies the B.S.C. A key observation is that L^+ and L^- solve the equation (they are classical solutions). Indeed

$$X_1 X_2 L^+(x) = -X_2 X_1 L^+(x),$$

$$X_1 X_2 L^-(x) = -X_2 X_1 L^-(x)$$

and

$$X_1 X_1 L^+(x) = X_2 X_2 L^+(x) = X_1 X_1 L^-(x) = X_2 X_2 L^-(x) = 0$$

so

$$\mathrm{div}_{\mathbb{H}}\big(a_i(\nabla_{\mathbb{H}} L^+)\big) = \sum_{i=1}^{2} X_i a_i(\nabla_{\mathbb{H}} L^+) = \sum_{i,j=1}^{2} \partial_{z_j} a_i(\nabla_{\mathbb{H}} L^+) X_i X_j L^+ = 0 \quad (3.48)$$

$$\mathrm{div}_{\mathbb{H}}\big(a_i(\nabla_{\mathbb{H}} L^-)\big) = \sum_{i=1}^{2} X_i a_i(\nabla_{\mathbb{H}} L^-) = \sum_{i,j=1}^{2} \partial_{z_j} a_i(\nabla_{\mathbb{H}} L^-) X_i X_j L^- = 0 \quad (3.49)$$

because $(\partial_{z_j} a_i)_{i,j}$ is a symmetric matrix while $(X_i X_j L^+)_{i,j}$ and $(X_i X_j L^-)_{i,j}$ are antisymmetric. Now D is a regular domain for the Dirichlet problem, therefore $u \in C(\overline{D})$ and it assumes the boundary datum ψ in the classical sense (we refer to [3] for boundary regularity) so that $u(y) = \psi(y)$ for all $y \in \partial D$.

This means that $L^-(y) = u(y) = L^+(y)$ and $L^-(x) \le u(x) \le L^+(x)$ for all $x \in \partial D$ and by the Comparison Principle

$$L^-(x) \le u(x) \le L^+(x) \quad \text{for all } x \in \overline{D}.$$

Now let $x \in \overline{D}$ and $y \in \partial D$. Then

$$u(x) - u(y) \le L^+(x) - u(y) = L^+(x) - L^+(y) \le M d(x, y) \tag{3.50}$$
$$u(y) - u(x) \le L^-(y) - u(x) = L^-(y) - L^-(x) \le M d(x, y) \tag{3.51}$$

where d is the Carnot-Carathéodory distance and by Remark 3.2 the Lipschitz constant has the dependences stated in the theorem.

Now we conclude by proving that

$$\sup_{x,y \in \overline{D}} \frac{|u(x) - u(y)|}{d(x, y)} = \sup_{\substack{x \in \overline{D} \\ y \in \partial D}} \frac{|u(x) - u(y)|}{d(x, y)}$$

in a separate lemma. $\qquad\square$

Lemma 3.7 *Let $u \in HW^{1,p}(D)$ be a weak solution of Eq. (3.1) in the open set D with smooth boundary datum. Then*

$$\sup_{x,y \in \overline{D}} \frac{|u(x) - u(y)|}{d(x, y)} = \sup_{\substack{x \in \overline{D} \\ y \in \partial D}} \frac{|u(x) - u(y)|}{d(x, y)}. \tag{3.52}$$

Proof Denote by M the right hand side of (3.52) and fix $x, y \in \overline{D}$, $x \ne y$. We will show that $|u(x) - u(y)| \le M d(x, y)$. Let $z = yx^{-1}$ and consider the function $u_z(w) = u(zw)$ which is defined in the set $D_z = \{w \mid zw \in D\}$ and it's a solution of Eq. (3.1) in this set. So both u and u_z are solutions of Eq. (3.1) in $D \cap D_z$ (note that this intersection is non empty because $x \in D$ and $y = zx \in D$).

For all $w \in \partial(D \cap D_z)$ we have

$$u(w) = u_z(w) + u(w) - u_z(w) \le u_z(w) + \sup_{w \subset \partial(D \cap D_z)} |u(w) - u_z(w)| = u_z(w) + C$$

but u and $u_z + C$ are both solutions of Eq. (3.1) in $D \cap D_z$, so by the Comparison Principle we get $u(w) \le u_z(w) + C$ for all $w \in D \cap D_z$.

Reversing the role of u and u_z we get that

$$|u(w) - u_z(w)| \le C \quad \text{for all } w \in D \cap D_z.$$

Now since u is continuous in $\overline{D}$ there exists $x_0 \in \partial(D \cap D_z)$ such that

$$|u(x) - u(y)| = |u(x) - u_z(x)| \le |u(x_0) - u_z(x_0)| = |u(x_0) - u(zx_0)| \tag{3.53}$$

and since either x_0 or zx_0 lie in ∂D we get that

$$|u(x_0) - u(zx_0)| = \frac{|u(x_0) - u(zx_0)|}{d(x_0, zx_0)} d(x_0, zx_0) \leq \sup_{\substack{x \in \overline{D} \\ y \in \partial D}} \frac{|u(x) - u(y)|}{d(x, y)} d(x_0, zx_0).$$

(3.54)

Since the Carnot-Carathéodory distance d is left invariant we have $d(x_0, zx_0) = d(x, y)$, so putting together (3.53) and (3.54) we get the result. $\square$

References

1. Giusti, E.: Direct Methods in the Calculus of Variations. World Scientific Publishing Co. Inc., River Edge (2003)
2. Miranda, M.: Un teorema di esistenza e unicità per il problema dell'area minima in *n* variabili. Ann. Scuola Norm. Sup. Pisa **3**(19), 233–249 (1965)
3. Trudinger, N.S., Wang, X.-J.: On the weak continuity of elliptic operators and applications to potential theory. Am. J. Math. **124**(2), 369–410 (2002)
4. Zhong, X.: Regularity for variational problems in the Heisenberg group. Preprint (2009)

Chapter 4
C^∞ Regularity for the Non-degenerate Equation

Abstract In this chapter we will study the regularity of weak solution to the non-degenerate p-Laplace equation (3.1)

$$\text{div}_{\mathbb{H}}\left(\left(\delta^2 + |\nabla_{\mathbb{H}} u|^2\right)^{\frac{p-2}{2}} \nabla_{\mathbb{H}} u\right) = 0 \quad \text{in } \Omega$$

for $1 < p < \infty$. We will use the results of the previous chapter, in particular that $u \in Lip(D)$ for a domain D satisfying (3.44) with

$$\|\nabla_{\mathbb{H}} u\|_{L^\infty(D)} \le M. \tag{4.1}$$

For this reason the results will be stated in terms of D instead of Ω, but since we are studying interior regularity and all the results are local, there is no loss of generality. We will use difference quotients to establish summability of derivatives in the vertical direction and then in the horizontal direction. Once we prove that Tu and $\nabla_{\mathbb{H}} u$ are in $HW_{loc}^{1,2}$ and satisfy certain linear equations we can apply well known regularity results to prove higher integrability of the solution u. We stress the fact the our estimates depend on the Lipschitz constant M and on the non degeneracy parameter δ, and blow up when δ goes to zero.

Keywords Non degenerate p-Laplace equation · Summability of derivatives · Interior regularity · Moser's iteration · Linear theory

4.1 Summability of Derivatives

The following theorems are an adaptation of the results of Capogna [1, 2] and Domokos [5].

Theorem 4.1 *Let $u \in HW^{1,p}(D)$, $1 < p < \infty$ be a weak solution of Eq. (3.1). Then $Tu \in L_{loc}^2(D)$. Moreover we have*

© The Author(s) 2015
D. Ricciotti, *p-Laplace Equation in the Heisenberg Group*,
SpringerBriefs in Mathematics, DOI 10.1007/978-3-319-23790-9_4

$$\|Tu\|^2_{L^2(B_r)} \leq \frac{C_{p,\delta,M}}{r^2} \|\nabla_{\mathbb{H}} u\|^2_{L^2(B_{2r})} \tag{4.2}$$

for every ball B_r such that the concentric ball $B_{2r} \subset D$.

Proof We use $\psi = \Delta^{\frac{1}{2}}_{T,-h}\varphi$, $\varphi \in C^\infty_0(\Omega)$ in (3.3). Since $\Delta^{\frac{1}{2}}_{T,h}$ commutes with X_i we have

$$0 = \int_D \sum_{i=1}^2 a_i(\nabla_{\mathbb{H}} u) X_i \left(\Delta^{\frac{1}{2}}_{T,-h}\varphi\right) dx = \int_D \sum_{i=1}^2 a_i(\nabla_{\mathbb{H}} u)\Delta^{\frac{1}{2}}_{T,-h}(X_i\varphi)\, dx$$

$$= \int_D \sum_{i=1}^2 \Delta^{\frac{1}{2}}_{T,h} a_i(\nabla_{\mathbb{H}} u) X_i\varphi\, dx \tag{4.3}$$

where we have used the discrete integration by parts formula in Proposition 2.4.

Now

$$\Delta^{\frac{1}{2}}_{T,h}\left(a_i(\nabla_{\mathbb{H}} u(x))\right) = \frac{a_i\left(\nabla_{\mathbb{H}} u(e^{hT}x)\right) - a_i\left(\nabla_{\mathbb{H}} u(x)\right)}{h^{\frac{1}{2}}}$$

$$= \frac{1}{h^{\frac{1}{2}}} \int_0^1 \frac{d}{dt} a_i\left(t\nabla_{\mathbb{H}} u(e^{hT}x) + (1-t)\nabla_{\mathbb{H}} u(x)\right) dt$$

$$= \frac{1}{h^{\frac{1}{2}}} \int_0^1 \sum_{j=1}^2 \partial_{z_j} a_i\left(t\nabla_{\mathbb{H}} u(e^{hT}x) + (1-t)\nabla_{\mathbb{H}} u(x)\right)$$

$$\times \left(X_j u(e^{hT}x) - X_j u(x)\right) dt$$

$$= \sum_{j=1}^2 \int_0^1 \partial_{z_j} a_i\left(t\nabla_{\mathbb{H}} u(e^{hT}x) + (1-t)\nabla_{\mathbb{H}} u(x)\right) dt\, \Delta^{\frac{1}{2}}_{T,h}(X_j u(x)). \tag{4.4}$$

Denoting $b_{i,j}(x) = \int_0^1 \partial_{z_j} a_i\left(t\nabla_{\mathbb{H}} u(e^{hT}x) + (1-t)\nabla_{\mathbb{H}} u(x)\right) dt$ Eq. (4.3) becomes

$$\int_D \sum_{i,j=1}^2 b_{i,j}(x)\, \Delta^{\frac{1}{2}}_{T,h}(X_j u(x))\, X_i\varphi = 0 \quad \text{for all } \varphi \in HW^{1,2}_0(\Omega) \tag{4.5}$$

and by (3.9)–(3.11) and (4.1), distinguishing the cases $1 < p < 2$ and $p \geq 2$, we get

$$|b_{i,j}(x)| \leq \overline{C}_{p,\delta,M} \tag{4.6}$$

$$\sum_{i,j=1}^2 b_{i,j}(x)\xi_i\xi_j \geq \overline{c}_{p,\delta,M}|\xi|^2 \tag{4.7}$$

where

$$\overline{C}_{p,\delta,M} = \begin{cases} C_p(\delta^2 + M^2)^{\frac{p-2}{2}} & \text{if } p \geq 2 \\ C_p\delta^{p-2} & \text{if } 1 < p < 2 \end{cases} \tag{4.8}$$

and

$$\overline{c}_{p,\delta,M} = \begin{cases} c_p\delta^{p-2} & \text{if } p \geq 2 \\ c_p(\delta^2 + M^2)^{\frac{p-2}{2}} & \text{if } 1 < p < 2. \end{cases} \tag{4.9}$$

Now use $\varphi = \xi^2 \Delta_{T,h}^{\frac{1}{2}} u$ as a test function in (4.5), where ξ is a cut-off function between the concentric balls $B_{\frac{3}{2}r}$ and B_R with $\frac{3}{2}r < R < 2r$ such that $|\nabla_{\mathbb{H}}\xi| \leq \frac{C}{r}$ (we refer to [4, 6–8] for the construction of such functions)

$$I = \int_D \sum_{i,j=1}^2 \xi^2 b_{i,j}(x) \, \Delta_{T,h}^{\frac{1}{2}} \left(X_j u(x)\right) X_i \left(\Delta_{T,h}^{\frac{1}{2}} u\right) dx$$

$$= -2 \int_D \sum_{i,j=1}^2 \xi \, b_{i,j}(x) \, \Delta_{T,h}^{\frac{1}{2}} \left(X_j u(x)\right) X_i \xi \, \Delta_{T,h}^{\frac{1}{2}} u \, dx = II. \tag{4.10}$$

Now since $\Delta_{T,h}^{\frac{1}{2}}$ commutes with X_i and using (4.7) we obtain

$$I \geq \overline{c}_{p,\delta,M} \int_D \xi^2 \, |\Delta_{T,h}^{\frac{1}{2}} \nabla_{\mathbb{H}} u|^2 dx, \tag{4.11}$$

while by (4.6) and Hölder's inequality

$$II \leq \overline{C}_{p,\delta,M} \int_D |\Delta_{T,h}^{\frac{1}{2}} \nabla_{\mathbb{H}} u| \, |\nabla_{\mathbb{H}} \xi| \, \xi \, |\Delta_{T,h}^{\frac{1}{2}} u| \, dx$$

$$\leq \overline{C}_{p,\delta,M} \left(\int_D \xi^2 |\Delta_{T,h}^{\frac{1}{2}} \nabla_{\mathbb{H}} u|^2 \, dx \right)^{\frac{1}{2}} \left(\int_D |\nabla_{\mathbb{H}}\xi|^2 |\Delta_{T,h}^{\frac{1}{2}} u|^2 dx \right)^{\frac{1}{2}}. \tag{4.12}$$

Now putting together (4.11) and (4.12), dividing by the first factor in the last inequality and squaring both sides we get

$$\int_D \xi^2 \, |\Delta_{T,h}^{\frac{1}{2}} \nabla_{\mathbb{H}} u|^2 dx \leq C_p \left(\frac{\delta^2 + M^2}{\delta^2} \right)^{|p-2|} \int_D |\nabla_{\mathbb{H}}\xi|^2 |\Delta_{T,h}^{\frac{1}{2}} u|^2 dx. \tag{4.13}$$

By the choice of ξ and Theorem 2.13 we get

$$\int_{B_{\frac{3}{2}r}} |\Delta_{T,h}^{\frac{1}{2}} \nabla_{\mathbb{H}} u|^2 dx \leq \frac{C_p}{r^2} \left(\frac{\delta^2 + M^2}{\delta^2} \right)^{|p-2|} \|\nabla_{\mathbb{H}} u\|_{L^2(B_R)}^2. \tag{4.14}$$

Using again Theorem 2.13 for the function $\Delta_{T,h}^{\frac{1}{2}} u \in HW^{1,2}(B_{\frac{3}{2}r})$ we have

$$\left\|\Delta_{T,-h}^{\frac{1}{2}}\left(\Delta_{T,h}^{\frac{1}{2}} u\right)\right\|_{L^2(B_r)} \leq C_p \left\|\Delta_{T,h}^{\frac{1}{2}} u\right\|_{HW^{1,2}(B_{\frac{3}{2}r})} \leq \frac{C_p}{r^2}\left(\frac{\delta^2+M^2}{\delta^2}\right)^{\frac{|p-2|}{}}\|\nabla_{\mathbb{H}} u\|_{L^2(B_R)}^2 .$$
(4.15)

By Remark 2.9

$$\left\|\Delta_{T,h}^{2,1} u\right\|_{L^2(B_r)} \leq \frac{C_p}{r^2}\left(\frac{\delta^2+M^2}{\delta^2}\right)^{|p-2|}\|\nabla_{\mathbb{H}} u\|_{L^2(B_R)}^2$$
(4.16)

and by Theorem 2.12

$$\left\|\Delta_{T,h}^{\frac{3}{4}} u\right\|_{L^2(B_r)} \leq \frac{C_p}{r^2}\left(\frac{\delta^2+M^2}{\delta^2}\right)^{|p-2|}\|\nabla_{\mathbb{H}} u\|_{L^2(B_R)}^2 .$$
(4.17)

Now repeating the same steps of (4.10)–(4.16) using $\varphi = \xi^2 \Delta_{T,h}^{\frac{3}{4}} u$ as a test function we get

$$\left\|\Delta_{T,h}^{2,\frac{5}{4}} u\right\|_{L^2(B_r)} \leq \frac{C_p}{r^2}\left(\frac{\delta^2+M^2}{\delta^2}\right)^{|p-2|}\|\nabla_{\mathbb{H}} u\|_{L^2(B_R)}^2$$
(4.18)

which by Theorem 2.12 yields

$$\left\|\Delta_{T,h} u\right\|_{L^2(B_r)} \leq \frac{C_p}{r^2}\left(\frac{\delta^2+M^2}{\delta^2}\right)^{|p-2|}\|\nabla_{\mathbb{H}} u\|_{L^2(B_R)}^2$$
(4.19)

so by Theorem 2.11 we get the result.   □

Theorem 4.2 Let $u \in HW^{1,p}(D)$, $1 < p < \infty$ be a weak solution of the non-degenerate p-Laplace equation (3.1). Then $Tu \in HW_{loc}^{1,2}(D)$. Moreover we have

$$\|\nabla_{\mathbb{H}} Tu\|_{L^2(B_r)}^2 \leq \frac{C_{p,\delta,M}}{r^2}\|Tu\|_{L^2(B_{2r})}^2 .$$
(4.20)

Proof Use $\psi = \Delta_{T,-h}\varphi$, where $\varphi \in C_0^\infty(\Omega)$ as a test function in (3.3). Using the fact that X_i and $\Delta_{T,-h}$ commute and the discrete integration by parts formula in Proposition (2.4) we have

$$0 = \int_D \sum_{i=1}^2 a_i(\nabla_{\mathbb{H}} u) X_i \left(\Delta_{T,-h}\varphi\right) \, dx = \int_D \sum_{i=1}^2 a_i(\nabla_{\mathbb{H}} u) \Delta_{T,-h}\left(X_i\varphi\right) \, dx$$

$$= \int_D \sum_{i=1}^2 \Delta_{T,h}\left(a_i(\nabla_{\mathbb{H}} u)\right) X_i\varphi \, dx = \int_D \sum_{i=1}^2 b_{i,j}(x) \Delta_{T,h}(X_j u) X_i\varphi \, dx,$$
(4.21)

where the $b_{i,j}(x) = \int_0^1 \partial_{z_j} a_i \left(t\nabla_{\mathbb{H}} u(e^{hT}x) + (1-t)\nabla_{\mathbb{H}} u(x) \right) \, dt$ satisfy

$$|b_{i,j}(x)| \leq \overline{C}_{p,\delta,M} \tag{4.22}$$

$$\sum_{i,j=1}^{2} b_{i,j}(x)\xi_i\xi_j \geq \overline{c}_{p,\delta,M}|\xi|^2 \tag{4.23}$$

with $\overline{C}_{p,\delta,M}$ and $\overline{c}_{p,\delta,M}$ as in (4.8) and (4.9). Choose $\varphi = \xi^2 \Delta_{T,h} u$ as a test function in (4.21) to get

$$I = \int_D \xi^2 \sum_{i=1}^{2} b_{i,j}(x)\Delta_{T,h}(X_j u)\, X_i\left(\Delta_{T,h}u\right)\, dx$$

$$= -2\int_D \xi \sum_{i=1}^{2} b_{i,j}(x)\, \Delta_{T,h}(X_j u)\, X_i\xi\, \Delta_{T,h}u\, dx = II. \tag{4.24}$$

Now using again $[T, X_i] = 0$ and (4.23) we obtain

$$I \geq \overline{c}_{p,\delta,M} \int_D \xi^2 \left|\Delta_{T,h}\nabla_{\mathbb{H}}u\right|^2 dx, \tag{4.25}$$

and by (4.22) and using Hölder's inequality with exponent 2

$$II \leq \overline{C}_{p,\delta,M} \int_D \xi\, |\Delta_{T,h}(X_j u)|\, |X_i\xi|\, |\Delta_{T,h}u|\, dx$$

$$\leq \overline{C}_{p,\delta,M} \left(\int_D \xi^2|\Delta_{T,h}\nabla_{\mathbb{H}}u|^2 dx\right)^{\frac{1}{2}} \left(\int_D |\nabla_{\mathbb{H}}\xi|^2|\Delta_{T,h}u|^2 dx\right)^{\frac{1}{2}}. \tag{4.26}$$

Now dividing the previous inequality by the first factor, and putting together (4.25)–(4.26) we get

$$\int_D \xi^2|\Delta_{T,h}\nabla_{\mathbb{H}}u|^2 dx \leq C_p \left(\frac{\delta^2 + M^2}{\delta^2}\right)^{|p-2|} \int_D |\nabla_{\mathbb{H}}\xi|^2|\Delta_{T,h}u|^2 dx. \tag{4.27}$$

Choosing now ξ to be a cut-off function between B_r and $B_{\frac{3}{2}r}$ satisfying $|\nabla_{\mathbb{H}}\xi| \leq \frac{c}{r}$ we get the result by Theorem 2.11. $\qquad\square$

Theorem 4.3 *Let $u \in HW^{1,p}(D)$, $1 < p < \infty$ be a weak solution of the non degenerate p-Laplace equation (3.1). Then $Tu \in L_{loc}^{\infty}(D)$. Moreover*

$$\|Tu\|_{L^\infty(B_r)} \le C_{p,\delta,M} \left(\fint_{B_{2r}} |Tu|^2 \mathrm{d}x \right)^{\frac{1}{2}} \tag{4.28}$$

for every B_r such that the concentric ball $B_{2r} \subset D$.

Proof Using $\psi = \Delta_{T,-h}\varphi$, where $\varphi \in C_0^\infty(\Omega)$ as a test function in (3.3) and proceeding as in the beginning of the proof of Theorem 4.2 we arrive at

$$\int_D \sum_{i=1}^2 b_{i,j}(x) \Delta_{T,h}(X_j u) \, X_i \varphi \, \mathrm{d}x = 0 \tag{4.29}$$

where the $b_{i,j}$ satisfy (4.22) and (4.23). Next use $\varphi = \xi^2 \left(\Delta_{T,h} u \right)^{2\alpha+1}$ as a test function in (4.29), where ξ is a cut-off function that will be chosen later and $\alpha \ge 0$. As we did in the proofs of the previous Theorems we get

$$\int_D \xi^2 |\Delta_{T,h}u|^{2\alpha} |\nabla_\mathbb{H} \Delta_{T,h}u|^2 \mathrm{d}x \le \frac{C_p}{(2\alpha+1)^2} \left(\frac{\delta^2 + M^2}{\delta^2} \right)^{|p-2|} \int_D |\nabla_\mathbb{H}\xi|^2 |\Delta_{T,h}u|^{2\alpha+2} \mathrm{d}x. \tag{4.30}$$

Now observe that

$$\int_D \left| \nabla_\mathbb{H} \left(\xi (\Delta_{T,h}u)^{\alpha+1} \right) \right|^2 \mathrm{d}x \le \int_D |\nabla_\mathbb{H}\xi|^2 |\Delta_{T,h}u|^{2\alpha+2} \mathrm{d}x$$

$$+ (\alpha+1)^2 \int_D \xi^2 |\Delta_{T,h}u|^{2\alpha} |\nabla_\mathbb{H}\Delta_{T,h}u|^2 \mathrm{d}x$$

$$\le C_{p,\delta,M} \|\nabla_\mathbb{H}\xi\|_{L^\infty}^2 \int_{\mathrm{supp}\xi} |\Delta_{T,h}u|^{2\alpha+2} \mathrm{d}x \tag{4.31}$$

where we have used (4.30). Next choose ξ_i, $i \ge 0$ to be cut-off functions between B_{r_i} and $B_{r_{i+1}}$ where $r_i = r + \frac{r}{2^{i+1}}$ and such that $|\nabla_\mathbb{H}\xi_i| \le \frac{C}{r_i - r_{i+1}}$. By Sobolev's embedding Theorem 2.8 with $q = 2$ we get

$$\left(\fint_{B_{r_i}} (\xi(\Delta_{T,h}u)^{\alpha+1})^{\frac{2Q}{Q-2}} \mathrm{d}x \right)^{\frac{Q-2}{Q}} \le C_{p,\delta,M} \left(\frac{r_i}{r_i - r_{i+1}} \right)^2 \fint_{B_{r_i}} |\Delta_{T,h}u|^{2\alpha+2} \mathrm{d}x \tag{4.32}$$

valid for all $\alpha \ge 0$. Now choose $\alpha = \alpha_i$ such that $\alpha_0 = 0$ and $2(\alpha_i + 1) = k^i$ for $i \ge 1$ with $k = \frac{Q}{Q-2}$. Raising (4.32) to the power $\frac{1}{k^i}$ we get

$$\left(\fint_{B_{r_{i+1}}} (\Delta_{T,h}u)^{k^{i+1}} \mathrm{d}x \right)^{\frac{1}{k^{i+1}}} \le \left(C_{p,\delta,M} 2^{2(i+2)} \right)^{\frac{1}{k^i}} \left(\fint_{B_{r_i}} |\Delta_{T,h}u|^{k^i} \mathrm{d}x \right)^{\frac{1}{k^i}}. \tag{4.33}$$

Iterating (4.33) by induction we obtain

$$
\begin{aligned}
\left(\fint_{B_{r_{n+1}}} (\Delta_{T,h} u)^{k^{n+1}} \, \mathrm{d}x \right)^{\frac{1}{k^{n+1}}} &\leq C_{p,\delta,M}^{\sum_{i=1}^{n} \frac{1}{k^i}} \prod_{i=1}^{n} 2^{\frac{2(i+2)}{k^i}} \left(\fint_{B_{\frac{3}{2}r}} (\Delta_{T,h} u)^2 \, \mathrm{d}x \right)^{\frac{1}{2}} \\
&\leq C_{p,\delta,M}^{\sum_{i=1}^{\infty} \frac{1}{k^i}} \prod_{i=1}^{\infty} 2^{\frac{2(i+2)}{k^i}} \left(\fint_{B_{\frac{3}{2}r}} (\Delta_{T,h} u)^2 \, \mathrm{d}x \right)^{\frac{1}{2}} \\
&\leq C_{p,\delta,M} \left(\fint_{B_{2r}} |Tu|^2 \mathrm{d}x \right)^{\frac{1}{2}}
\end{aligned}
\tag{4.34}
$$

since $\log \prod_{k=1}^{\infty} 2^{\frac{2(i+2)}{k^i}} = 2 \log 2 \sum_{i=1}^{\infty} \frac{i+2}{k^i}$ is a finite constant and we have used Theorem 2.11 which also yields

$$
\left(\fint_{B_{r_{n+1}}} |Tu|^{k^{n+1}} \mathrm{d}x \right)^{\frac{1}{k^{n+1}}} \leq C_{p,\delta,M} \left(\fint_{B_{2r}} |Tu|^2 \mathrm{d}x \right)^{\frac{1}{2}}.
\tag{4.35}
$$

Since k^n tends to infinity when n tends to infinity, passing to limit we get the result. $\square$

Theorem 4.4 *Let $u \in HW^{1,p}(D)$, $1 < p < \infty$ be a weak solution of the non-degenerate p-Laplace equation (3.1). Then $\nabla_{\mathbb{H}} u \in HW^{1,2}_{loc}(D)$. Moreover*

$$
\|\nabla_{\mathbb{H}}^2 u\|^2_{L^2(B_r)} \leq \frac{C_{p,\delta,M}}{r^2} \|\nabla_{\mathbb{H}} u\|^2_{L^2(B_{2r})} + C_{p,\delta,M} \|Tu\|^2_{HW^{1,2}(B_{2r})}
\tag{4.36}
$$

for every ball B_r such that the concentric ball $B_{2r} \subset D$.

Proof Use $\psi = \Delta_{X_1,-h} \varphi$, where $\varphi \in C_0^{\infty}(\Omega)$ as a test function in (3.3). Using Lemma 2.2 and the discrete integration by parts formula in Proposition 2.4 we get

$$
\begin{aligned}
0 &= \int_D \sum_{i=1}^{2} a_i(\nabla_{\mathbb{H}} u) X_i \left(\Delta_{X_1,-h} \varphi \right) \, \mathrm{d}x \\
&= \int_D \sum_{i=1}^{2} a_i(\nabla_{\mathbb{H}} u) \Delta_{X_1,-h} (X_i \varphi) \, \mathrm{d}x - \int_D a_2(\nabla_{\mathbb{H}} u) T \varphi(\mathrm{e}^{hX_1} x) \, \mathrm{d}x \\
&= \int_D \sum_{i=1}^{2} \Delta_{X_1,h} (a_i(\nabla_{\mathbb{H}} u)) X_i \varphi \, \mathrm{d}x - \int_D a_2(\nabla_{\mathbb{H}} u) T \varphi(\mathrm{e}^{hX_1} x) \, \mathrm{d}x.
\end{aligned}
\tag{4.37}
$$

Now

$$
\begin{aligned}
\Delta_{X_1,h}\Big(a_i(\nabla_{\mathbb{H}} u(x))\Big) &= \frac{a_i\left(\nabla_{\mathbb{H}} u(e^{hX_1}x)\right) - a_i\left(\nabla_{\mathbb{H}} u(x)\right)}{h} \\
&= \frac{1}{h}\int_0^1 \frac{d}{dt} a_i\left(t\nabla_{\mathbb{H}} u(e^{hX_1}x) + (1-t)\nabla_{\mathbb{H}} u(x)\right) dt \\
&= \frac{1}{h}\int_0^1 \sum_{j=1}^2 \partial_{z_j} a_i\left(t\nabla_{\mathbb{H}} u(e^{hX_1}x) + (1-t)\nabla_{\mathbb{H}} u(x)\right) \\
&\qquad \times \left(X_j u(e^{hX_1}x) - X_j u(x)\right) dt \\
&= \sum_{j=1}^2 \int_0^1 \partial_{z_j} a_i\left(t\nabla_{\mathbb{H}} u(e^{hX_1}x) + (1-t)\nabla_{\mathbb{H}} u(x)\right) dt\, \Delta_{X_1,h}\left(X_j u(x)\right).
\end{aligned}
$$

(4.38)

Denoting with $b_{i,j}(x) = \int_0^1 \partial_{z_j} a_i\left(t\nabla_{\mathbb{H}} u(e^{hX_1}x) + (1-t)\nabla_{\mathbb{H}} u(x)\right) dt$ from (4.37) and (4.38) we get

$$
\int_D \sum_{i,j=1}^2 b_{i,j}(x)\Delta_{X_1,h}\left(X_j u\right) X_i \varphi\, dx - \int_D a_2(\nabla_{\mathbb{H}} u)T\varphi(e^{hX_1}x)\, dx = 0 \qquad (4.39)
$$

and by (3.9)–(3.11) and (4.1)

$$
|b_{i,j}(x)| \leq \overline{C}_{p,\delta,M} \qquad (4.40)
$$

$$
\sum_{i,j=1}^2 b_{i,j}(x)\xi_i\xi_j \geq \overline{c}_{p,\delta,M}|\xi|^2 \qquad (4.41)
$$

with $\overline{C}_{p,\delta,M}$ and $\overline{c}_{p,\delta,M}$ as in (4.8) and (4.9). Use $\varphi = \xi^2 \Delta_{X_1,h} u$ as a test function in (4.39), where ξ is a cut-off function between B_r and $B_{\frac{3}{2}r}$ such that $|\nabla_{\mathbb{H}}\xi| \leq \frac{C}{r}$ and $|T\xi| \leq \frac{C}{r^2}$ (it is sufficient to build a cut-off function between B_1 and B_2 as in [6, 7] and use dilations $\delta_{\frac{1}{r}}$ keeping in mind that T is homogeneous of degree 2 with respect to $\delta_{\frac{1}{r}}$). We get

$$
\int_D \xi^2 \sum_{i,j=1}^2 b_{i,j}(x)\Delta_{X_1,h}\left(X_j u\right) X_i\left(\Delta_{X_1,h} u\right) dx
$$

$$
+ 2\int_D \xi \sum_{i,j=1}^2 b_{i,j}(x)\Delta_{X_1,h}\left(X_j u\right) X_i\xi\, \Delta_{X_1,h} u\, dx
$$

$$
- \int_D a_2(\nabla_{\mathbb{H}} u)T\varphi(e^{hX_1}x)\, dx = 0. \qquad (4.42)
$$

Using Lemma 2.2 we get

$$
\begin{aligned}
0 = & \int_D \xi^2 \sum_{i,j=1}^2 b_{i,j}(x)\Delta_{X_1,h}\left(X_j u\right)\Delta_{X_1,h}\left(X_i u\right)\, dx \\
& + \int_D \xi^2 \sum_{j=1}^2 b_{2,j}(x)Tu(e^{hX_1}x)\Delta_{X_1,h}\left(X_j u\right)\, dx \\
& + 2\int_D \xi \sum_{i,j=1}^2 b_{i,j}(x)\Delta_{X_1,h}\left(X_j u\right)X_i\xi\,\Delta_{X_1,h}u\, dx - \int_D a_2(\nabla_{\mathbb{H}}u)T\varphi(e^{hX_1}x)\, dx \\
= & \; I + II + III + IV.
\end{aligned}
$$

(4.43)

Now by (4.41)

$$
I \geq \overline{c}_{p,\delta,M}\int_D \xi^2 |\Delta_{X_1,h}\left(\nabla_{\mathbb{H}}u\right)|^2 dx.
$$

(4.44)

Using (4.40) and Young's inequality with exponent 2 introducing a parameter $\varepsilon > 0$ to be suitably chosen later we get

$$
\begin{aligned}
II \leq & \; \overline{C}_{p,\delta,M}\int_D \xi^2\,|Tu(e^{hX_1}x)||\Delta_{X_1,h}\nabla_{\mathbb{H}}u|\, dx \\
\leq & \; \varepsilon\overline{C}_{p,\delta,M}\int_D \xi^2|\Delta_{X_1,h}\nabla_{\mathbb{H}}u|^2 dx + \frac{\overline{C}_{p,\delta,M}}{\varepsilon}\int_D \xi^2\,|Tu(e^{hX_1}x)|^2 dx \\
\leq & \; \varepsilon\overline{C}_{p,\delta,M}\int_D \xi^2|\Delta_{X_1,h}\nabla_{\mathbb{H}}u|^2 dx + \frac{\overline{C}_{p,\delta,M}}{\varepsilon}\int_{B_{2r}} |Tu(x)|^2 dx
\end{aligned}
$$

(4.45)

provided h is small enough. Analogously

$$
\begin{aligned}
III \leq & \; \overline{C}_{p,\delta,M}\int_D \xi\,|\Delta_{X_1,h}\nabla_{\mathbb{H}}u|\,|\nabla_{\mathbb{H}}\xi|\,|\Delta_{X_1,h}u|\, dx \\
\leq & \; \varepsilon\overline{C}_{p,\delta,M}\int_D \xi^2\,|\Delta_{X_1,h}\nabla_{\mathbb{H}}u|^2 dx + \frac{\overline{C}_{p,\delta,M}}{\varepsilon}\int_D |\nabla_{\mathbb{H}}\xi|^2\,|\Delta_{X_1,h}u|^2 dx.
\end{aligned}
$$

(4.46)

We are left with

$$
\begin{aligned}
IV = & \; 2\int_D a_2(\nabla_{\mathbb{H}}u)\xi(e^{hX_1}x)T\xi(e^{hX_1}x)\Delta_{X_1,h}u(e^{hX_1}x)\, dx \\
& + \int_D a_2(\nabla_{\mathbb{H}}u)\xi^2(e^{hX_1}x)T\Delta_{X_1,h}u(e^{hX_1}x)\, dx \\
= & \; IV_1 + IV_2.
\end{aligned}
$$

(4.47)

Now since ξ has compact support, provided h is small enough we have

$$IV_1 \le \overline{C}'_{p,\delta,M} \int_D |\xi(x)|\,|T\xi(x)|\,|\Delta_{X_1,h}u(x)|\mathrm{d}x \qquad (4.48)$$

where

$$\overline{C}'_{p,\delta,M} = \begin{cases} C_p(\delta^2 + M^2)^{\frac{p-1}{2}} & \text{if } p \ge 2 \\ C_p\delta^{p-1} & \text{if } 1 < p < 2. \end{cases} \qquad (4.49)$$

Analogously we get

$$IV_2 \le \overline{C}'_{p,\delta,M} \int_D \xi^2 |\Delta_{X_1,h}Tu|\,\mathrm{d}x. \qquad (4.50)$$

Putting together (4.44)–(4.50) and choosing $\varepsilon = \frac{1}{2}\left(\frac{\delta^2}{\delta^2+M^2}\right)^{\frac{|p-2|}{2}}$ we get

$$\int_{B_r} \xi^2 |\Delta_{X_1,h}(\nabla_{\mathbb{H}}u)|^2\mathrm{d}x \le C_{p,\delta,M}\left(\int_{B_{2r}} |Tu|^2\mathrm{d}x + \int_{B_{2r}} |\nabla_{\mathbb{H}}Tu|^2\mathrm{d}x \right.$$
$$\left. + (\|\xi T\xi\|_{L^\infty(D)} + \|\nabla_{\mathbb{H}}\xi\|^2_{L^\infty(D)})\int_{B_{\frac{3}{2}r}} |\Delta_{X_1,h}u|^2\mathrm{d}x\right).$$
$$(4.51)$$

Now using Theorem 2.11 and repeating the same steps for the test function $\psi = \Delta_{X_2,-h}\varphi$ we get the result. $\qquad\square$

Lemma 4.1 *The functions $v_1 = X_1u$, $v_2 = X_2u$ and $v_3 = Tu$ are weak solutions respectively of the following equations (in D):*

$$\sum_{i=1}^{2} X_i\left(\sum_{j=1}^{2} \partial_{z_j}a_i(\nabla_{\mathbb{H}}u)X_jv_1\right) + \sum_{i=1}^{2} X_i\left(\partial_{z_2}a_i(\nabla_{\mathbb{H}}u)Tu\right) + T\left(a_2(\nabla_{\mathbb{H}}u)\right) = 0$$
$$(4.52)$$

$$\sum_{i=1}^{2} X_i\left(\sum_{j=1}^{2} \partial_{z_j}a_i(\nabla_{\mathbb{H}}u)X_jv_2\right) - \sum_{i=1}^{2} X_i\left(\partial_{z_1}a_i(\nabla_{\mathbb{H}}u)Tu\right) - T\left(a_1(\nabla_{\mathbb{H}}u)\right) = 0$$
$$(4.53)$$

$$\sum_{i=1}^{2} X_i\left(\sum_{j=1}^{2} \partial_{z_j}a_i(\nabla_{\mathbb{H}}u)X_jv_3\right) = 0. \qquad (4.54)$$

Proof We start proving (4.52). The key idea is to consider $\varphi \in C_0^\infty(D)$ and use $\psi = X_1\varphi$ as a test function and then integrate by parts. The only problem is that the horizontal vector field do not commute so some terms involving the vertical vector field T will appear.

The function u satisfies

$$\int_D \sum_{i=1}^2 a_i(\nabla_{\mathbb{H}} u) X_i \psi \, dx = 0 \quad \text{for all } \psi \in C_0^\infty(D). \tag{4.55}$$

With our choice of ψ we get

$$\int_D \sum_{i=1}^2 a_i(\nabla_{\mathbb{H}} u) X_i X_1 \varphi \, dx = 0. \tag{4.56}$$

To see what kind of equation v_1 satisfies we need to integrate by parts with respect to X_1. Keeping in mind the commutation relation $X_1 X_2 - X_2 X_1 = T$ we get

$$\int_D \sum_{i=1}^2 a_i(\nabla_{\mathbb{H}} u) X_1 X_i \varphi \, dx - \int_D a_2(\nabla_{\mathbb{H}} u) T \varphi \, dx = 0. \tag{4.57}$$

By Theorem 4.4 $v_1 \in HW^{1,2}(D)$ so in the first integral we can integrate by parts with respect to X_1 and get

$$\int_D \sum_{i=1}^2 a_i(\nabla_{\mathbb{H}} u) X_1 X_i \varphi \, dx = -\int_D \sum_{i=1}^2 X_1 \left(a_i(\nabla_{\mathbb{H}} u) \right) X_i \varphi \, dx$$

$$= -\int_D \sum_{i=1}^2 \sum_{j=1}^2 \partial_{z_j} a_i(\nabla_{\mathbb{H}} u) X_1 X_j u X_i \varphi \, dx$$

$$= -\int_D \sum_{i,j=1}^2 \partial_{z_j} a_i(\nabla_{\mathbb{H}} u) X_j v_1 X_i \varphi \, dx$$

$$\quad - \int_D \sum_{i=1}^2 \partial_{z_2} a_i(\nabla_{\mathbb{H}} u) T u X_i \varphi \, dx. \tag{4.58}$$

By (4.57) and (4.58) the function v_1 satisfies

$$\int_D \sum_{i=1}^2 \left(\sum_{i=1}^2 \partial_{z_j} a_i(\nabla_{\mathbb{H}} u) X_j v_1 \right) X_i \varphi + \int_D \sum_{i=1}^2 \partial_{z_2} a_i(\nabla_{\mathbb{H}} u) T u X_i \varphi + a_2(\nabla_{\mathbb{H}} u) T \varphi = 0 \tag{4.59}$$

which is the weak formulation of (4.52).

To prove (4.53) use $\psi = X_2 \varphi$ as a test function in (4.55), the commutator relation $X_1 X_2 = X_2 X_1 + T$ and integration by parts to get

$$0 = \int_D \sum_{i=1}^2 a_i(\nabla_\mathbb{H} u) X_i X_2 \varphi \, dx$$

$$= \int_D \sum_{i=1}^2 a_i(\nabla_\mathbb{H} u) X_2 X_i \varphi \, dx + \int_D a_1(\nabla_\mathbb{H} u) T \varphi \, dx$$

$$= -\int_D \sum_{i=1}^2 X_2 \left(a_i(\nabla_\mathbb{H} u) \right) X_i \varphi \, dx + \int_D a_1(\nabla_\mathbb{H} u) T \varphi \, dx$$

$$= -\int_D \sum_{i=1}^2 \left(\sum_{j=1}^2 \partial_{z_j} a_i(\nabla_\mathbb{H} u) X_2 X_j u \right) X_i \varphi \, dx + \int_D a_1(\nabla_\mathbb{H} u) T \varphi \, dx$$

$$= -\int_D \sum_{i=1}^2 \left(\sum_{j=1}^2 \partial_{z_j} a_i(\nabla_\mathbb{H} u) X_j v_2 \right) X_\varphi \, dx + \int_D \sum_{i=1}^2 \partial_{z_1} a_i(\nabla_\mathbb{H} u) T u X_i \varphi \, dx$$

$$+ \int_D a_1(\nabla_\mathbb{H} u) T \varphi \, dx$$

which is the weak formulation of (4.53).

To prove (4.54) use $\psi = T\varphi$ as a test function in (4.55). This time X_i and T commute, so we can exchange their order and integrate by parts by virtue of the regularity result in Theorem 4.2 getting:

$$0 = -\int_D \sum_{i=1}^2 a_i(\nabla_\mathbb{H} u) X_i T \varphi \, dx = -\int_D \sum_{i=1}^2 a_i(\nabla_\mathbb{H} u) T X_i \varphi \, dx$$

$$= \int_D \sum_{i=1}^2 T \left(a_i(\nabla_\mathbb{H} u) \right) X_i \varphi \, dx = \int_D \sum_{i=1}^2 \left(\sum_{j=1}^2 \partial_{z_j} a_i(\nabla_\mathbb{H} u) T X_j u \right) X_i \varphi \, dx$$

$$= \int_D \sum_{i=1}^2 \left(\sum_{j=1}^2 \partial_{z_j} a_i(\nabla_\mathbb{H} u) X_j v_3 \right) X_i \varphi \, dx \tag{4.60}$$

which is the weak formulation of (4.54).   □

Remark 4.1 Equations (4.52)–(4.54) can be respectively written in the following form

$$\sum_{i=1}^2 X_i \left(\sum_{j=1}^2 w_{i,j}(x) X_j v_1 + w_i(x) \right) = w(x) \tag{4.61}$$

$$\sum_{i=1}^2 X_i \left(\sum_{j=1}^2 w_{i,j}(x) X_j v_2 + w_i'(x) \right) = w'(x) \tag{4.62}$$

$$\sum_{i,j=1}^{2} X_i \left(w_{i,j}(x) X_j v_3 \right) = 0 \tag{4.63}$$

where

$$w_{i,j}(x) = \partial_{z_j} a_i(\nabla_{\mathbb{H}} u(x)), \tag{4.64}$$

$$w_i(x) = \partial_{z_2} a_i(\nabla_{\mathbb{H}} u(x)), \tag{4.65}$$

$$w(x) = -T \left(a_2(\nabla_{\mathbb{H}} u(x)) \right), \tag{4.66}$$

$$w_i'(x) = -\partial_{z_1} a_i(\nabla_{\mathbb{H}} u(x)), \tag{4.67}$$

and

$$w'(x) = T \left(a_1(\nabla_{\mathbb{H}} u(x)) \right). \tag{4.68}$$

These equations are linear and the regularity theory regarding their solution is well known, and it requires the use of Morrey and Campanato spaces which were introduced in Chap. 2.

4.2 Linear Theory

Definition 4.1 (*Potential*) Let $f \in L^1_{loc}(\Omega)$. Define

$$V(f)(x) = \int_{\Omega} \Gamma(x, y) f(y) \, dy \tag{4.69}$$

where Γ is the fundamental solution for the operator $L = \sum_{j=1}^{2} X_j^2$.

We refer to [1] for more details and properties of the fundamental solution of sub-Laplacians on Carnot groups.

Theorem 4.5 *Let K be a compact set. Then there exist constants $C, R_0 > 0$ such that for all $x \in K$, $0 < r < R_0$*

$$C^{-1} \frac{d(x, y)^2}{\left| B_{d(x,y)}(x) \right|} \leq \Gamma(x, y) \leq C \frac{d(x, y)^2}{\left| B_{d(x,y)}(x) \right|} \tag{4.70}$$

$$\left| \nabla_{\mathbb{H}, y} \Gamma(x, y) \right| \leq C \frac{d(x, y)}{\left| B_{d(x,y)}(x) \right|}. \tag{4.71}$$

Theorem 4.6 *Let Ω be a bounded open subset of $\mathbb{H}$ and $f \in L^1_{loc}(\Omega)$ such that for a certain $\lambda \in]0, 1[$ it holds*

$$\fint_{\Omega_r(x_0)} |f| \, dx \leq C r^{\lambda - 2} \quad \text{for all } x_0 \in \Omega, \ 0 < r < \text{diam}(\Omega). \tag{4.72}$$

Then

$$\fint_{\Omega_r(x_0)} |\nabla_{\mathbb{H}} V(f)|^2 dx \le C r^{2(\lambda-1)}. \tag{4.73}$$

Proof Let $x_0 \in \Omega$ and $x \in B_r(x_0)$. Using the estimates on the fundamental solution provided in Theorem 4.5 we get

$$|\nabla_{\mathbb{H}} V(f)(x)| \le \int_\Omega |\nabla_{\mathbb{H}} \Gamma(x,y)| \, |f(y)| \, dy \le C \int_\Omega \frac{d(x,y)}{|B_{d(x,y)}(x)|} |f(y)| \, dy$$

$$\le C \left(\int_{\Omega_{2r}(x_0)} \frac{d(x,y)}{|B_{d(x,y)}(x)|} |f(y)| \, dy + \int_{\Omega \setminus \Omega_{2r}(x_0)} \frac{d(x,y)}{|B_{d(x,y)}(x)|} |f(y)| \, dy \right)$$

$$= I(x) + II(x). \tag{4.74}$$

Let $\sigma > 0$ be a real number that will be chosen later and use Hölder's inequality to get

$$I(x) = \int_{\Omega_{2r}(x_0)} \frac{d(x,y)^{\frac{\sigma}{2}+1-\frac{\sigma}{2}}}{|B_{d(x,y)}(x)|} |f(y)| \, dy$$

$$\le \left(\int_{\Omega_{2r}(x_0)} \frac{d(x,y)^\sigma}{|B_{d(x,y)}(x)|} |f(y)| \, dy \right)^{\frac{1}{2}} \left(\int_{\Omega_{2r}(x_0)} \frac{d(x,y)^{2-\sigma}}{|B_{d(x,y)}(x)|} |f(y)| \, dy \right)^{\frac{1}{2}}$$

$$= I_1^{\frac{1}{2}}(x) I_2^{\frac{1}{2}}(x). \tag{4.75}$$

Now since $B_{2r}(x_0) \subset B_{4r}(x)$ we have

$$I_2(x) \le \int_{\Omega_{4r}(x)} \frac{d(x,y)^{2-\sigma}}{|B_{d(x,y)}(x)|} |f(y)| \, dy = \sum_{k=-2}^\infty \int_{A_k \cap \Omega} \frac{d(x,y)^{2-\sigma}}{|B_{d(x,y)}(x)|} |f(y)| \, dy \tag{4.76}$$

where $A_k = B_{\frac{r}{2^k}}(x) \setminus B_{\frac{r}{2^{k+1}}}(x)$. If $y \in A_k$ then $d(x,y) \le \frac{r}{2^k}$ and $|B_{d(x,y)}(x)| \ge |B_{\frac{r}{2^{k+1}}}(x)|$, therefore using (4.72) we get

$$I_2(x) \le \sum_{k=-2}^\infty \left(\frac{r}{2^k} \right)^{2-\sigma} \frac{1}{|B_{\frac{r}{2^{k+1}}}(x)|} \int_{A_k \cap \Omega} |f(y)| \, dy$$

$$\le \sum_{k=-2}^\infty \left(\frac{r}{2^k} \right)^{2-\sigma} \frac{|B_{\frac{r}{2^k}}(x)|}{|B_{\frac{r}{2^{k+1}}}(x)|} \fint_{\Omega_{\frac{r}{2^k}}(x)} |f(y)| \, dy$$

$$\le C \sum_{k=-2}^\infty \left(\frac{r}{2^k} \right)^{\lambda-\sigma} \le C r^{\lambda-\sigma} \tag{4.77}$$

if we choose $0 < \sigma < \lambda$. From (4.75) and (4.77) we get

$$\int_{\Omega_r(x_0)} I^2(x)\,dx \le Cr^{\lambda-\sigma} \int_{\Omega_r(x_0)} \int_{\Omega_{2r}(x_0)} \frac{d(x,y)^\sigma}{|B_{d(x,y)}(x)|} |f(y)|\,dy\,dx$$

$$= Cr^{\lambda-\sigma} \int_{\Omega_{2r}(x_0)} |f(y)| \int_{\Omega_r(x_0)} \frac{d(x,y)^\sigma}{|B_{d(x,y)}(x)|}\,dx\,dy. \qquad (4.78)$$

Now proceeding as in (4.76)–(4.77) we get

$$\int_{\Omega_r(x_0)} \frac{d(x,y)^\sigma}{|B_{d(x,y)}(x)|}\,dx \le Cr^\sigma \qquad (4.79)$$

therefore

$$\int_{\Omega_r(x_0)} I^2(x)\,dx \le Cr^\lambda |B_r(x_0)| \fint_{\Omega_{2r}(x_0)} |f(y)|\,dy \le Cr^{2(\lambda-1)} |B_r(x_0)|. \qquad (4.80)$$

Now in order to estimate $II(x)$ extend f to be zero outside Ω and observe that

$$\Omega \setminus B_{2r}(x_0) \subset B_R(x_0) \setminus B_{2r}(x_0) \subset \bigcup_{k=0}^{k_0} A_k$$

where $R = \operatorname{diam}(\Omega)$, $A_k = B_{\frac{R}{2^k}}(x) \setminus B_{\frac{R}{2^{k+1}}}(x)$ and $k_0 = \left[\log_2 \frac{R}{r} + 1\right]$, so

$$II(x) \le \sum_{k=0}^{k_0} \int_{A_k \cap \Omega} \frac{d(x,y)}{|B_{d(x,y)}(x)|} |f(y)|\,dy$$

$$\le \sum_{k=0}^{k_0} \frac{R}{2^k} \frac{|B_{\frac{R}{2^k}}(x)|}{|B_{\frac{R}{2^{k+1}}}(x)|} \fint_{\Omega_{\frac{R}{2^k}}(x)} |f(y)|\,dy$$

$$\le \sum_{k=0}^{k_0} \frac{R}{2^k} \left(\frac{R}{2^k}\right)^{\lambda-2} \le Cr^{\lambda-1}. \qquad (4.81)$$

From (4.74), (4.80) and (4.81) we obtain the desired result.   □

Theorem 4.7 *Let Ω be a bounded open subset of $\mathbb{H}$ and $v \in HW^{1,2}(\Omega)$ be a weak solution of a linear equation*

$$\sum_{i=1}^{2} X_i \left(\sum_{j=1}^{2} w_{i,j}(x) X_j v + w_i(x) \right) = w(x) \qquad (4.82)$$

whose coefficients satisfy the following assumptions

$$w_{i,j} \in L^\infty_{loc}(\Omega) \quad and \quad W = \left(w_{i,j}\right)_{i,j} \quad is \ a \ positive \ definite \ matrix; \qquad (4.83)$$

$$w_i \in M^{2,\lambda}_{loc}(\Omega); \qquad\qquad\qquad\qquad\qquad\qquad\qquad\qquad\qquad (4.84)$$

$$\fint_{B_r} |w| \, dx \le Cr^{\lambda-2} \quad and \quad w \in L^{\frac{Q}{Q+2}}(\Omega). \qquad\qquad (4.85)$$

Then there exists $\overline{\lambda} \in]0,1]$, $\overline{\lambda} \le \lambda$ such that

$$|\nabla_{\mathbb{H}} v| \in M^{2,\overline{\lambda}}_{loc}(\Omega) \quad and \quad v \in \Gamma^{\overline{\lambda}}(\Omega). \qquad\qquad (4.86)$$

Proof From Theorem 4.6 we have that $|\nabla_{\mathbb{H}} V(w)| \in M^{2,\lambda}(\Omega)$. Moreover $V(w) = \Gamma \star w$ satisfies

$$\sum_{i=1}^{2} X_i^2 V(w) = w$$

because Γ is the fundamental solution for the sub-Laplacian $L = \sum_{i=1}^{2} X_i^2$. Therefore

$$\sum_{i=1}^{2} X_i \left(w_{i,j} X_j u + w_i\right) = \sum_{i=1}^{2} X_i^2 V(w)$$

so Eq. (4.82) is equivalent to the following

$$\sum_{i=1}^{2} X_i \left(w_{i,j} X_j u + w_i + X_i V(w)\right) = 0$$

where $w_i + X_i V(w) \in M^{2,\lambda}(\Omega)$. For this reason it is sufficient to prove the Theorem for $w = 0$.

Consider $x \in B_R(x_0)$ and $0 < r < R$. By Lax-Milgram's Theorem there exists a unique solution of the problem

$$\begin{cases} \sum_{i=1}^{2} X_i(w_{i,j} X_j H) = 0 & in \ B_r(x) \\ H - u \in HW^{1,2}_0(B_r(x)). \end{cases} \qquad\qquad (4.87)$$

Moreover the solution H belongs to $\Gamma^\alpha(B_r(x))$ for $0 < \alpha < 1$ (see [3]). Next consider a cut-off function ξ between $B_{\frac{3}{4}r}$ and $B_{\frac{r}{2}}$ such that $|\nabla_{\mathbb{H}}\xi| \le \frac{C}{r}$ and use $\varphi(y) = (H(y) - H(x))\xi^2(y)$ as a test function in the weak formulation of (4.87) to get

$$\int_{\Omega} \xi^2 X_j H\, X_i H\, dy \leq 2 \int_{\Omega} \xi(y)\, w_{i,j}(y) X_j H(y)\, X_i \xi(y)\, (H(y) - H(x))\, dy$$

$$\leq C \int_{B_{\frac{3}{4}r}} |\nabla_{\mathbb{H}} H|\, |\nabla_{\mathbb{H}} \xi|\, \xi\, |H - H(y)|\, dy$$

$$\leq C \left(\int_{B_{\frac{3}{4}r}} |\nabla_{\mathbb{H}} H|^2 \xi^2\, dy \right)^{\frac{1}{2}} \left(\int_{B_{\frac{3}{4}r}} |\nabla_{\mathbb{H}} \xi|^2\, |H - H(y)|^2\, dy \right)^{\frac{1}{2}}.$$

$$(4.88)$$

From assumption (4.83) we have

$$\int_{\Omega} \xi^2 X_j H\, X_i H\, dy \geq \int_{B_{\frac{r}{2}}} |\nabla_{\mathbb{H}} H|^2 dy \qquad (4.89)$$

and together with (4.88) we get to

$$\fint_{B_{\frac{r}{2}}} |\nabla_{\mathbb{H}} H|^2\, dy \leq \frac{C}{r^2} \fint_{B_{\frac{3}{4}r}} |H - H(x)|^2 dy \leq C r^{2(\alpha-1)} \qquad (4.90)$$

and therefore $|\nabla_{\mathbb{H}} H| \in M_{loc}^{2,\alpha}(\Omega)$.

Now observe that $U = u - H$ is a solution of Eq. (4.82) in $B_r(x)$ and since $u - H \in HW_0^{1,2}(B_r(x))$ we can use it as a test function in the weak formulation of Eq. (4.82) to get

$$\fint_{B_r(x)} |\nabla_{\mathbb{H}} (u - H)|^2 dx \leq \fint_{B_r(x)} |w_i|\, |\nabla_{\mathbb{H}} u - H|\, dx. \qquad (4.91)$$

Using Young's inequality with exponent 2 we have

$$\fint_{B_r(x)} |\nabla_{\mathbb{H}} (u - H)|^2 dx \leq \fint_{B_r(x)} |w_i|^2 dx \leq C r^{2(\lambda-1)} \qquad (4.92)$$

thanks to (4.84). Therefore we have

$$\fint_{B_r(x)} |\nabla_{\mathbb{H}} u|^2 dx \leq C r^{2(\lambda-1)} + \fint_{B_r(x)} |\nabla_{\mathbb{H}} H|^2 dx \leq C(r^{2(\lambda-1)} + r^{2(\alpha-1)}) \qquad (4.93)$$

which means that $|\nabla_{\mathbb{H}} u| \in M^{2, min\{\lambda, \alpha\}}(B_r(x_0))$. $\qquad \square$

Remark 4.2 Following [9] it can be proved that $\bar{\lambda}$ can be chosen to be exactly λ.

Theorem 4.8 *Let $u \in HW^{1,p}(\Omega)$, $1 < p < \infty$ be a weak solution of the non degenerate p-Laplace equation (3.1). Then $\nabla_{\mathbb{H}} u \in \Gamma^{\lambda}(\Omega)$ and $Tu \in \Gamma^{\lambda}(\Omega)$ for any $\lambda \in]0, 1[$.*

Proof Since we are interested in interior regularity we can assume $u \in Lip(\Omega)$. $X_1 u$, $X_2 u$ and $Tu \in HW_{loc}^{1,2}(\Omega)$ satisfy (4.61)–(4.63). Since we are considering solutions to the non-degenerate equation (3.1), the coefficients of (4.63) satisfy the assumptions (4.83) of Theorem 4.7 for every $\lambda \in]0, 1[$. Therefore we get $|\nabla_\mathbb{H} Tu| \in M_{loc}^{2,\lambda}(\Omega)$ and $Tu \in \Gamma^\lambda(\Omega)$, for every $\lambda \in]0, 1[$.

Since $\|\nabla_\mathbb{H} u\|_{L^\infty(\Omega)} \leq M$ using (3.11) we get

$$\fint_{B_r} |w_i|^2 dx \leq \overline{C}_{p,\delta,M} \leq \overline{C}_{p,\delta,M} r^{2(\lambda-1)} \tag{4.94}$$

and

$$\fint_{B_r} |w_i'|^2 dx \leq \overline{C}_{p,\delta,M} \leq \overline{C}_{p,\delta,M} r^{2(\lambda-1)} \tag{4.95}$$

for $r < 1$ and for all $\lambda \in]0, 1[$, where $\overline{C}_{p,\delta,M}$ is the same constant of (4.8). Therefore the coefficients w_i and w_i' from (4.61) and (4.62) satisfy (4.84).

Now since $|\nabla_\mathbb{H} Tu| \in M_{loc}^{2,\lambda}(\Omega)$

$$\fint_{B_r} |w| \, dx \leq C \left(\fint_{B_r} |\nabla_\mathbb{H} Tu|^2 dx \right)^{\frac{1}{2}} \leq Cr^{\lambda-1} \leq Cr^{\lambda-2} \tag{4.96}$$

for $r < 1$. Therefore it is now possible to apply Theorem 4.7 to obtain that $X_1 u$, $X_2 u \in \Gamma^\lambda(\Omega)$. $\square$

Once we have the result of Theorem 4.8, the interior higher regularity follows by a well known bootstrap argument. For the sake of completeness we state the results and refer to [1] and references therein for detailed proofs.

Proposition 4.1 *Let $v \in HW^{1,2}(\Omega) \cap \Gamma^\alpha(\Omega)$ be a weak solution of Eq. (4.82) with $w_{i,j}$, $w_i \in \Gamma^\alpha(\Omega)$ and $w \in L^\infty(\Omega)$. Then $v \in \Gamma^{1,\alpha}(\Omega)$.*

Corollary 4.1 *Let $u \in HW^{1,p}(\Omega)$, $1 < p < \infty$ be a weak solution of the non-degenerate p-Laplace equation (3.1). Then $Tu \in \Gamma^{1,\alpha}(\Omega)$ and $\nabla_\mathbb{H} u \in \Gamma^{1,\alpha}(\Omega)$.*

Proof Consider Eq. (4.63) satisfied by $v_3 = Tu$. Since $a_i \in C^\infty(\Omega)$ and $\nabla_\mathbb{H} u \in \Gamma^\alpha(\Omega)$ then $w_{i,j} = \partial_{z_j} a_i(\nabla_\mathbb{H} u)$ satisfy the assumption of Proposition 4.1, therefore $Tu \in \Gamma^{1,\alpha}(\Omega)$. Now consider Eqs. (4.61) and (4.62) satisfied respectively by $v_1 = X_1 u$ and $v_2 = X_2 u$. Arguing as above we see that $w_{i,j}$, w_i and w_i' belong to $\Gamma^\alpha(\Omega)$. Moreover

$$|w| \leq \sum_{j=1}^{2} |\partial_{z_j} a_2(\nabla_\mathbb{H} u)| \, |\nabla_\mathbb{H} Tu| \leq C|\nabla_\mathbb{H} Tu| \in L_{loc}^\infty(\Omega)$$

because $Tu \in \Gamma^\alpha(\Omega)$. We can apply Proposition 4.1 and get $X_1 u \in \Gamma^{1,\alpha}(\Omega)$. Completely analogous is the proof for $X_2 u$. $\square$

Theorem 4.9 *Let $v \in \Gamma^{2,\alpha}(\Omega)$ be a solution to the equation*

$$\sum_{i,j=1}^{2} \omega_{i,j} X_i X_j v + \omega = 0 \qquad (4.97)$$

with $\omega_{i,j}, \omega \in \Gamma^{k,\alpha}(\Omega)$. Then $v \in \Gamma^{k+2,\alpha}(\Omega)$.

Corollary 4.2 *Let $u \in HW^{1,p}(\Omega)$, $1 < p < \infty$ be a weak solution of the non-degenerate p-Laplace equation (3.1). Then $u \in C^{\infty}(\Omega)$.*

Proof By Corollary 4.1 $u \in \Gamma^{2,\alpha}(\Omega)$ we can differentiate Eq. (3.1) to get

$$\sum_{i,j=1}^{2} \partial_{z_j} a_i(\nabla_{\mathbb{H}} u) X_i X_j u = 0. \qquad (4.98)$$

Since $a_i \in C^{\infty}(\Omega)$ and $\nabla_{\mathbb{H}} u \in \Gamma^{1,\alpha}(\Omega)$ from Theorem 4.9 we get $u \in \Gamma^{3,\alpha}(\Omega)$. This means $\nabla_{\mathbb{H}} u \in \Gamma^{2,\alpha}(\Omega)$ so the coefficients of Eq. (4.98) belong to $\Gamma^{2,\alpha}(\Omega)$, and again by Theorem 4.9 we get $u \in \Gamma^{4,\alpha}(\Omega)$. Iterating this argument we get eventually $u \in C^{\infty}(\Omega)$. $\qquad\square$

References

1. Capogna, L.: Regularity of quasi-linear equations in the Heisenberg group. Comm. Pure Appl. Math. **50**(9), 867–889 (1997)
2. Capogna, L.: Regularity for quasilinear equations and 1-quasiconformal maps in Carnot groups. Math. Ann. **313**(2), 263–295 (1999)
3. Capogna, L., Danielli, D., Garofalo, N.: An embedding theorem and the Harnack inequality for nonlinear subelliptic equations. Comm. Partial Differ. Equ. **18**(9–10), 1765–1794 (1993)
4. Citti, G., Garofalo, N., Lanconelli, E.: Harnack's inequality for sum of squares of vector fields plus a potential. Amer. J. Math. **115**(3), 699–734 (1993)
5. Domokos, A.: Differentiability of solutions for the non-degenerate p-Laplacian in the Heisenberg group. J. Differ. Equ. **204**(2), 439–470 (2004)
6. Franchi, B., Serapioni, R., Serra Cassano, F.: Approximation and imbedding theorems for weighted Sobolev spaces associated with Lipschitz continuous vector fields. Boll. Un. Mat. Ital. B (7) **11**(1), 83–117 (1997)
7. Garofalo, N., Nhieu, D.-M.: Lipschitz continuity, global smooth approximations and extension theorems for Sobolev functions in Carnot-Carathéodory spaces. J. Anal. Math. **74**, 67–97 (1998)
8. Monti, R., Serra Cassano, F.: Surface measures in Carnot-Carathéodory spaces. Calc. Var. Partial Differ. Equ. **13**(3), 339–376 (2001)
9. Morrey Jr, C.B.: Second order elliptic equations in several variables and Hölder continuity. Math. Z **72**, 146–164 (1959/1960)

Chapter 5
Lipschitz Regularity

Abstract In this chapter we will prove the Lipschitz regularity of solution to the degenerate p-Laplace equation for $1 < p < \infty$ following (Zhong, Regularity for variational problems in the Heisenberg group, 2009 [1]). To achieve this we will try to obtain estimates independent of the non degeneracy parameter δ when dealing with solutions to the non-degenerate equation, and then pass to the limit for $\delta \to 0$.

Keywords Lipschitz regularity · Caccioppoli estimates · Moser's iteration

5.1 Caccioppoli Type Estimates

By the Hilbert-Haar theory in Sect. 3.3 we are able to produce Lipschitz solutions in domains satisfying a certain convexity-type property, namely condition (3.34). In this section we assume that D is such a domain, so we have that there exists a constant $M > 0$ such that

$$\|\nabla_{\mathbb{H}} u\|_{L^\infty(D)} \leq M. \tag{5.1}$$

By the results of Chap. 4 we have

$$\nabla_{\mathbb{H}} u \in HW^{1,2}_{loc}(D) \tag{5.2}$$

$$Tu \in HW^{1,2}_{loc}(D) \cap L^\infty_{loc}(D). \tag{5.3}$$

With these regularity results we can avoid the use of difference quotients in the following proofs.

We will try to procede analogously to the Euclidean case. The aim is to use a suitable version of Moser's iteration to gain the estimate on $\|\nabla_{\mathbb{H}} u\|_{L^\infty(D)}$. To do this we need to produce some Caccioppoli type estimates to control the Tu term. The next results are adapted from [1].

© The Author(s) 2015
D. Ricciotti, *p-Laplace Equation in the Heisenberg Group*,
SpringerBriefs in Mathematics, DOI 10.1007/978-3-319-23790-9_5

Lemma 5.1 *Let $\alpha \geq 0$ and $\xi \in C_0^\infty(D)$. Then we have the following Caccioppoli type estimate*

$$\int_D \xi^{\alpha+2} \left(\delta^2 + |\nabla_\mathbb{H} u|^2\right)^{\frac{p-2}{2}} |Tu|^\alpha |\nabla_\mathbb{H} Tu|^2 dx$$

$$\leq C_p \int_D \xi^\alpha |\nabla_\mathbb{H} \xi|^2 \left(\delta^2 + |\nabla_\mathbb{H} u|^2\right)^{\frac{p-2}{2}} |Tu|^{\alpha+2} dx.$$

Proof By the regularity result (5.3) we have $\varphi = \xi^{\alpha+2}|Tu|^{\alpha+1} \in HW_0^{1,2}(D)$. Using it as a test function in (4.54) we get

$$(\alpha+1) \int_D \sum_{i=1}^2 \left(\sum_{j=1}^2 \partial_{z_j} a_i(\nabla_\mathbb{H} u) X_j Tu\right) |Tu|^\alpha X_i Tu \, \xi^{\alpha+2} \, dx$$

$$= -(\alpha+2) \int_D \sum_{i=1}^2 \left(\sum_{j=1}^2 \partial_{z_j} a_i(\nabla_\mathbb{H} u) X_j Tu\right) X_i \xi \, \xi^{\alpha+1} |Tu|^{\alpha+1} dx. \qquad (5.4)$$

Denote by I the left hand side of the previous equation and by II the right one. Using the ellipticity property (3.10) we have that

$$I \geq (\alpha+1)c_p \int_D \xi^{\alpha+2} \left(\delta^2 + |\nabla_\mathbb{H} u|^2\right)^{\frac{p-2}{2}} |Tu|^\alpha |\nabla_\mathbb{H} Tu|^2 \, dx \qquad (5.5)$$

and using (3.11) we can estimate the right hand side

$$II \leq (\alpha+2)C_p \int_D \left(\delta^2 + |\nabla_\mathbb{H} u|^2\right)^{\frac{p-2}{2}} |\nabla_\mathbb{H} Tu||\nabla_\mathbb{H} \xi| \, \xi^{\alpha+1} |Tu|^{\alpha+1} dx$$

$$\leq (\alpha+1)C_p \left(\int_D \xi^{\alpha+2} \left(\delta^2 + |\nabla_\mathbb{H} u|^2\right)^{\frac{p-2}{2}} |Tu|^\alpha |\nabla_\mathbb{H} Tu|^2 \, dx\right)^{1/2}$$

$$\times \left(\int_D \xi^\alpha |\nabla_\mathbb{H} \xi|^2 \left(\delta^2 + |\nabla_\mathbb{H} u|^2\right)^{\frac{p-2}{2}} |Tu|^{\alpha+2} dx\right)^{1/2} \qquad (5.6)$$

where we have used Hölder's inequality with exponent 2.

Now putting together (5.5) and (5.6), squaring both sides and dividing by the first factor of the right hand side we get

$$\int_D \xi^{\alpha+2} \left(\delta^2 + |\nabla_\mathbb{H} u|^2\right)^{\frac{p-2}{2}} |Tu|^\alpha |\nabla_\mathbb{H} Tu|^2 dx$$

$$\leq C_p \int_D |\nabla_\mathbb{H} \xi|^2 \left(\delta^2 + |\nabla_\mathbb{H} u|^2\right)^{\frac{p-2}{2}} |Tu|^{\alpha+2} dx.$$

$\square$

Lemma 5.2 *Let $\alpha \geq 0$ and $\xi \in C_0^\infty(D)$. Then*

$$\int_D \xi^2 \left(\delta^2 + |\nabla_{\mathbb{H}} u|^2\right)^{\frac{p-2+\alpha}{2}} |\nabla_{\mathbb{H}}^2 u|^2 dx \leq C_p (\alpha + 1)^3$$

$$\times \int_D (|\nabla_{\mathbb{H}} \xi|^2 + \xi |T\xi|) \left(\delta^2 + |\nabla_{\mathbb{H}} u|^2\right)^{\frac{p+\alpha}{2}} dx$$

$$+ C_p (\alpha+1)^4 \int_D \xi^2 \left(\delta^2 + |\nabla_{\mathbb{H}} u|^2\right)^{\frac{p-2+\alpha}{2}} |Tu|^2 dx.$$

Proof Using $\varphi = \xi^2 \left(\delta^2 + |\nabla_{\mathbb{H}} u|^2\right)^{\frac{\alpha}{2}} X_1 u$ as a test function in the weak formulation of equation (4.52) we get

$$I_1 = \int_D \xi^2 \sum_{i,j=1}^2 \partial_{z_j} a_i (\nabla_{\mathbb{H}} u) X_j X_1 u \ X_i \left(\left(\delta^2 + |\nabla_{\mathbb{H}} u|^2\right)^{\frac{\alpha}{2}} X_1 u\right) dx$$

$$= -2 \int_D \sum_{i,j=1}^2 \partial_{z_j} a_i (\nabla_{\mathbb{H}} u) X_j X_1 u \ X_i \xi \ \xi \left(\delta^2 + |\nabla_{\mathbb{H}} u|^2\right)^{\frac{\alpha}{2}} X_1 u dx$$

$$- \int_D \sum_{i=1}^2 \partial_{z_2} a_i (\nabla_{\mathbb{H}} u) Tu \ X_i \left(\xi^2 \left(\delta^2 + |\nabla_{\mathbb{H}} u|^2\right)^{\frac{\alpha}{2}} X_1 u\right) dx$$

$$- \int_D a_2 (\nabla_{\mathbb{H}} u) T \left(\xi^2 \left(\delta^2 + |\nabla_{\mathbb{H}} u|^2\right)^{\frac{\alpha}{2}} X_1 u\right) dx$$

$$= I_2 + II + III. \tag{5.7}$$

First we will estimate the right hand side of (5.7).

By property (3.11) and using Young's inequality with exponent 2 introducing a parameter $\varepsilon > 0$ that will be chosen later we can estimate I_2 in the following way

$$I_2 \leq C_p \int_D \left(\delta^2 + |\nabla_{\mathbb{H}} u|^2\right)^{\frac{p-1+\alpha}{2}} \xi |\nabla_{\mathbb{H}} \xi| |\nabla_{\mathbb{H}} X_1 u| dx$$

$$\leq \varepsilon C_p \int_D \xi^2 \left(\delta^2 + |\nabla_{\mathbb{H}} u|^2\right)^{\frac{p-2+\alpha}{2}} |\nabla_{\mathbb{H}} X_1 u|^2 dx + \frac{C_p}{\varepsilon} \int_D \left(\delta^2 + |\nabla_{\mathbb{H}} u|^2\right)^{\frac{p+\alpha}{2}} |\nabla_{\mathbb{H}} \xi|^2 dx$$

$$\leq \varepsilon C_p \int_D \xi^2 \left(\delta^2 + |\nabla_{\mathbb{H}} u|^2\right)^{\frac{p-2+\alpha}{2}} |\nabla_{\mathbb{H}}^2 u|^2 dx + \frac{C_p}{\varepsilon} \int_D \left(\delta^2 + |\nabla_{\mathbb{H}} u|^2\right)^{\frac{p+\alpha}{2}} |\nabla_{\mathbb{H}} \xi|^2 dx. \tag{5.8}$$

Now calculating derivatives we get

$$II = 2 \int_D \sum_{i=1}^2 \partial_{z_2} a_i (\nabla_{\mathbb{H}} u) Tu \ X_i \xi \ \xi \left(\delta^2 + |\nabla_{\mathbb{H}} u|^2\right)^{\frac{\alpha}{2}} X_1 u \ dx$$

$$+ \alpha \int_D \xi^2 \sum_{i=1}^2 \partial_{z_2} a_i (\nabla_{\mathbb{H}} u) Tu \left(\delta^2 + |\nabla_{\mathbb{H}} u|^2\right)^{\frac{\alpha-2}{2}} \langle X_i \nabla_{\mathbb{H}} u, \nabla_{\mathbb{H}} u \rangle X_1 u \ dx$$

$$+ \int_D \xi^2 \sum_{i=1}^{2} \partial_{z_2} a_i (\nabla_{\mathbb{H}} u) T u \ \left(\delta^2 + |\nabla_{\mathbb{H}} u|^2\right)^{\frac{\alpha}{2}} X_i X_1 u \ \mathrm{d}x$$

$$= II_1 + II_2 + II_3. \tag{5.9}$$

From (3.11) and Young's inequality with a parameter $\varepsilon > 0$ to be suitably chosen later we have

$$II_1 \le C_p \int_D \xi |\nabla_{\mathbb{H}} \xi| \left(\delta^2 + |\nabla_{\mathbb{H}} u|^2\right)^{\frac{p-1+\alpha}{2}} |T u| \ \mathrm{d}x$$

$$\le \varepsilon C_p \int_D \xi^2 \left(\delta^2 + |\nabla_{\mathbb{H}} u|^2\right)^{\frac{p-2+\alpha}{2}} |T u|^2 \ \mathrm{d}x + \frac{C_p}{\varepsilon} \int_D |\nabla_{\mathbb{H}} \xi|^2 \left(\delta^2 + |\nabla_{\mathbb{H}} u|^2\right)^{\frac{p+\alpha}{2}} \ \mathrm{d}x$$

$$\le \varepsilon C_p \int_D \xi^2 \left(\delta^2 + |\nabla_{\mathbb{H}} u|^2\right)^{\frac{p-2+\alpha}{2}} |\nabla_{\mathbb{H}}^2 u|^2 \ \mathrm{d}x + \frac{C_p}{\varepsilon} \int_D |\nabla_{\mathbb{H}} \xi|^2 \left(\delta^2 + |\nabla_{\mathbb{H}} u|^2\right)^{\frac{p+\alpha}{2}} \ \mathrm{d}x$$

where we have used $|T u| \le 2|\nabla_{\mathbb{H}}^2 u|$.
Analogously

$$II_2 \le \alpha C_p \int_D \xi^2 \left(\delta^2 + |\nabla_{\mathbb{H}} u|^2\right)^{\frac{p-2+\alpha}{2}} |T u| \ |X_i \nabla_{\mathbb{H}} u| \ \mathrm{d}x$$

$$\le \varepsilon \alpha C_p \int_D \xi^2 \left(\delta^2 + |\nabla_{\mathbb{H}} u|^2\right)^{\frac{p-2+\alpha}{2}} |\nabla_{\mathbb{H}}^2 u|^2 \ \mathrm{d}x$$

$$+ \alpha \frac{C_p}{\varepsilon} \int_D \xi^2 \left(\delta^2 + |\nabla_{\mathbb{H}} u|^2\right)^{\frac{p-2+\alpha}{2}} |T u|^2 \ \mathrm{d}x.$$

$$II_3 \le \varepsilon C_p \int_D \xi^2 \left(\delta^2 + |\nabla_{\mathbb{H}} u|^2\right)^{\frac{p-2+\alpha}{2}} |\nabla_{\mathbb{H}}^2 u|^2 \ \mathrm{d}x$$

$$+ \frac{C_p}{\varepsilon} \int_D \xi^2 \left(\delta^2 + |\nabla_{\mathbb{H}} u|^2\right)^{\frac{p-2+\alpha}{2}} |T u|^2 \ \mathrm{d}x.$$

We will estimate III in the following way

$$III = 2 \int_D a_2(\nabla_{\mathbb{H}} u) \xi \ T\xi \left(\delta^2 + |\nabla_{\mathbb{H}} u|^2\right)^{\frac{\alpha}{2}} X_1 u \ \mathrm{d}x$$

$$+ \alpha \int_D \xi^2 a_2(\nabla_{\mathbb{H}} u) \left(\delta^2 + |\nabla_{\mathbb{H}} u|^2\right)^{\frac{\alpha-2}{2}} \langle T \nabla_{\mathbb{H}} u, \nabla_{\mathbb{H}} u \rangle X_1 u \ \mathrm{d}x$$

$$+ \int_D \xi^2 a_2(\nabla_{\mathbb{H}} u) \left(\delta^2 + |X_1 u|^2\right)^{\frac{\alpha}{2}} T X_1 u \ \mathrm{d}x$$

$$= III_1 + III_2 + III_3. \tag{5.10}$$

Using (3.9) we get

$$III_1 \le C_p \int_D \left(\delta^2 + |\nabla_{\mathbb{H}} u|^2\right)^{\frac{p+\alpha}{2}} \xi |T\xi| \ \mathrm{d}x. \tag{5.11}$$

Now to produce the estimates we want for III_2 we need to integrate by parts

$$III_2 = \alpha \int_D \xi^2 a_2(\nabla_{\mathbb{H}} u) \left(\delta^2 + |\nabla_{\mathbb{H}} u|^2\right)^{\frac{\alpha-2}{2}} \sum_{k=1}^2 X_k T u \, X_k u \, X_1 u \, dx$$

$$= -\alpha \int_D \sum_{k=1}^2 X_k \left(\xi^2 a_2(\nabla_{\mathbb{H}} u) \left(\delta^2 + |\nabla_{\mathbb{H}} u|^2\right)^{\frac{\alpha-2}{2}} X_k u \, X_1 u\right) T u \, dx$$

$$= -2\alpha \int_D \xi \sum_{k=1}^2 X_k \xi \, a_2(\nabla_{\mathbb{H}} u) \left(\delta^2 + |\nabla_{\mathbb{H}} u|^2\right)^{\frac{\alpha-2}{2}} X_k u \, X_1 u \, T u \, dx$$

$$- \alpha \int_D \xi^2 \sum_{j,k=1}^2 \partial_{z_j} a_2(\nabla_{\mathbb{H}} u) X_k X_j u \left(\delta^2 + |\nabla_{\mathbb{H}} u|^2\right)^{\frac{\alpha-2}{2}} X_k u \, X_1 u \, T u \, dx$$

$$- \alpha(\alpha - 2) \int_D \xi^2 a_2(\nabla_{\mathbb{H}} u) \left(\delta^2 + |\nabla_{\mathbb{H}} u|^2\right)^{\frac{\alpha-4}{2}} \langle X_k \nabla_{\mathbb{H}} u, \nabla_{\mathbb{H}} u \rangle X_k u \, X_1 u \, T u \, dx$$

$$- \alpha \int_D \xi^2 a_2(\nabla_{\mathbb{H}} u) \left(\delta^2 + |\nabla_{\mathbb{H}} u|^2\right)^{\frac{\alpha-2}{2}} X_k X_k u \, X_1 u \, T u \, dx$$

$$- \alpha \int_D \xi^2 a_2(\nabla_{\mathbb{H}} u) \left(\delta^2 + |\nabla_{\mathbb{H}} u|^2\right)^{\frac{\alpha-2}{2}} X_k u \, X_k X_1 u \, T u \, dx$$

$$= III_{2,1} + III_{2,2} + III_{2,3} + III_{2,4} + III_{2,5}. \tag{5.12}$$

Using (3.9) and Young's inequality with exponent 2 with a parameter $\varepsilon > 0$ to be chosen later we get

$$III_{2,1} \leq \alpha C_p \int_D \left(\delta^2 + |X_1 u|^2\right)^{\frac{p-1+\alpha}{2}} \xi |\nabla_{\mathbb{H}} \xi| |T u| dx$$

$$\leq \varepsilon \alpha C_p \int_D \xi^2 \left(\delta^2 + |\nabla_{\mathbb{H}} u|^2\right)^{\frac{p-2+\alpha}{2}} |T u|^2 dx$$

$$+ \alpha \frac{C_p}{\varepsilon} \int_D |\nabla_{\mathbb{H}} \xi|^2 \left(\delta^2 + |\nabla_{\mathbb{H}} u|^2\right)^{\frac{p+\alpha}{2}} dx$$

$$\leq \varepsilon \alpha C_p \int_D \xi^2 \left(\delta^2 + |\nabla_{\mathbb{H}} u|^2\right)^{\frac{p-2+\alpha}{2}} |\nabla_{\mathbb{H}}^2 u|^2 dx$$

$$+ \alpha \frac{C_p}{\varepsilon} \int_D |\nabla_{\mathbb{H}} \xi|^2 \left(\delta^2 + |\nabla_{\mathbb{H}} u|^2\right)^{\frac{p+\alpha}{2}} dx \tag{5.13}$$

where in the last inequality we have used the fact that $|T u| \leq 2|\nabla_{\mathbb{H}}^2 u|$.

Now using (3.11)

$$
\begin{aligned}
III_{2,2} &\leq \alpha C_p \int_D \xi^2 \left(\delta^2 + |\nabla_{\mathbb{H}} u|^2\right)^{\frac{p-2+\alpha}{2}} |\nabla_{\mathbb{H}}^2 u||Tu| \, dx \\
&\leq \varepsilon \alpha C_p \int_D \xi^2 \left(\delta^2 + |\nabla_{\mathbb{H}} u|^2\right)^{\frac{p-2+\alpha}{2}} |\nabla_{\mathbb{H}}^2 u|^2 dx \\
&\quad + \alpha \frac{C_p}{\varepsilon} \int_D \xi^2 \left(\delta^2 + |\nabla_{\mathbb{H}} u|^2\right)^{\frac{p-2+\alpha}{2}} |Tu|^2 dx.
\end{aligned}
\tag{5.14}
$$

Analogously

$$
\begin{aligned}
III_{2,3} &\leq \alpha(\alpha - 2) C_p \int_D \xi^2 \left(\delta^2 + |\nabla_{\mathbb{H}} u|^2\right)^{\frac{p-2+\alpha}{2}} |\nabla_{\mathbb{H}}^2 u||Tu| \, dx \\
&\leq \alpha(\alpha - 2)\varepsilon C_p \int_D \xi^2 \left(\delta^2 + |\nabla_{\mathbb{H}} u|^2\right)^{\frac{p-2+\alpha}{2}} |\nabla_{\mathbb{H}}^2 u|^2 dx \\
&\quad + \alpha(\alpha - 2)\frac{C_p}{\varepsilon} \int_D \xi^2 \left(\delta^2 + |\nabla_{\mathbb{H}} u|^2\right)^{\frac{p-2+\alpha}{2}} |Tu|^2 dx.
\end{aligned}
$$

$$
\begin{aligned}
III_{2,4} &\leq \alpha \int_D \xi^2 \left(\delta^2 + |\nabla_{\mathbb{H}} u|^2\right)^{\frac{p-2+\alpha}{2}} |\nabla_{\mathbb{H}}^2 u||Tu| \, dx \\
&\leq \varepsilon \alpha C_p \int_D \xi^2 \left(\delta^2 + |\nabla_{\mathbb{H}} u|^2\right)^{\frac{p-2+\alpha}{2}} |\nabla_{\mathbb{H}}^2 u|^2 dx \\
&\quad + \alpha \frac{C_p}{\varepsilon} \int_D \xi^2 \left(\delta^2 + |\nabla_{\mathbb{H}} u|^2\right)^{\frac{p-2+\alpha}{2}} |Tu|^2 dx.
\end{aligned}
$$

$$
\begin{aligned}
III_{2,5} &\leq \varepsilon \alpha C_p \int_D \xi^2 \left(\delta^2 + |\nabla_{\mathbb{H}} u|^2\right)^{\frac{p-2+\alpha}{2}} |\nabla_{\mathbb{H}}^2 u|^2 dx \\
&\quad + \alpha \frac{C_p}{\varepsilon} \int_D \xi^2 \left(\delta^2 + |\nabla_{\mathbb{H}} u|^2\right)^{\frac{p-2+\alpha}{2}} |Tu|^2 dx.
\end{aligned}
\tag{5.15}
$$

Completely similar is the estimate for III_3; first we integrate by parts producing more integrals $III_{3,1}, III_{3,2}, III_{3,3}$,

$$
\begin{aligned}
III_3 &= \int_D \xi^2 a_2(\nabla_{\mathbb{H}} u) \left(\delta^2 + |\nabla_{\mathbb{H}} u|^2\right)^{\frac{\alpha}{2}} X_1 Tu \, dx \\
&= -\int_D X_1 \left(\xi^2 a_2(\nabla_{\mathbb{H}} u) \left(\delta^2 + |\nabla_{\mathbb{H}} u|^2\right)^{\frac{\alpha}{2}} \right) Tu \, dx \\
&= -2 \int_D \xi \, X_1 \xi \, a_2(\nabla_{\mathbb{H}} u) \left(\delta^2 + |\nabla_{\mathbb{H}} u|^2\right)^{\frac{\alpha}{2}} Tu \, dx \\
&\quad - \int_D \xi^2 \sum_{j=1}^{2} \partial_{z_j} a_2(\nabla_{\mathbb{H}} u) X_1 X_j u \left(\delta^2 + |\nabla_{\mathbb{H}} u|^2\right)^{\frac{\alpha}{2}} Tu \, dx \\
&\quad - \alpha \int_D \xi^2 a_2(\nabla_{\mathbb{H}} u) \left(\delta^2 + |\nabla_{\mathbb{H}} u|^2\right)^{\frac{\alpha-2}{2}} \langle X_1 \nabla_{\mathbb{H}} u, \nabla_{\mathbb{H}} u \rangle Tu \, dx \\
&= III_{3,1} + III_{3,2} + III_{3,3}.
\end{aligned}
\tag{5.16}
$$

Now

$$III_{3,1} \leq \varepsilon C_p \int_D \xi^2 \left(\delta^2 + |\nabla_{\mathbb{H}} u|^2\right)^{\frac{p-2+\alpha}{2}} |\nabla_{\mathbb{H}}^2 u|^2 dx$$
$$+ \frac{C_p}{\varepsilon} \int_D |\nabla_{\mathbb{H}} \xi|^2 \left(\delta^2 + |\nabla_{\mathbb{H}} u|^2\right)^{\frac{p+\alpha}{2}} dx.$$

$$III_{3,2} \leq \varepsilon C_p \int_D \xi^2 \left(\delta^2 + |\nabla_{\mathbb{H}} u|^2\right)^{\frac{p-2+\alpha}{2}} |\nabla_{\mathbb{H}}^2 u|^2 dx$$
$$+ \frac{C_p}{\varepsilon} \int_D \xi^2 \left(\delta^2 + |\nabla_{\mathbb{H}} u|^2\right)^{\frac{p-2+\alpha}{2}} |Tu|^2 dx.$$

$$III_{3,3} \leq \alpha\varepsilon C_p \int_D \xi^2 \left(\delta^2 + |\nabla_{\mathbb{H}} u|^2\right)^{\frac{p-2+\alpha}{2}} |\nabla_{\mathbb{H}}^2 u|^2 dx$$
$$+ \alpha \frac{C_p}{\varepsilon} \int_D \xi^2 \left(\delta^2 + |\nabla_{\mathbb{H}} u|^2\right)^{\frac{p-2+\alpha}{2}} |Tu|^2 dx. \tag{5.17}$$

Putting together (5.8)–(5.17) we have

$$I_1 \leq \varepsilon(\alpha+1)^2 C_p \int_D \xi^2 \left(\delta^2 + |\nabla_{\mathbb{H}} u|^2\right)^{\frac{p-2+\alpha}{2}} |\nabla_{\mathbb{H}}^2 u|^2 dx$$
$$+ \frac{(\alpha+1)C_p}{\varepsilon} \int_D \left(|\nabla_{\mathbb{H}} \xi|^2 + \xi|T\xi|\right) \left(\delta^2 + |\nabla_{\mathbb{H}} u|^2\right)^{\frac{p+\alpha}{2}} dx$$
$$+ \frac{(\alpha+1)^2 C_p}{\varepsilon} \int_D \xi^2 \left(\delta^2 + |\nabla_{\mathbb{H}} u|^2\right)^{\frac{p-2+\alpha}{2}} |Tu|^2 dx. \tag{5.18}$$

Now in a completely similar way, using $\varphi = \xi^2 \left(\delta^2 + |\nabla_{\mathbb{H}} u|^2\right)^{\frac{\alpha}{2}} X_2 u$ in the weak formulation of (4.53) we get

$$J_1 = \int_D \xi^2 \sum_{i,j=1}^{2} \partial_{z_j} a_i(\nabla_{\mathbb{H}} u) X_j X_2 u \ X_i \left(\left(\delta^2 + |\nabla_{\mathbb{H}} u|^2\right)^{\frac{\alpha}{2}} X_2 u\right) dx$$
$$\leq \varepsilon(\alpha+1)^2 C_p \int_D \xi^2 \left(\delta^2 + |\nabla_{\mathbb{H}} u|^2\right)^{\frac{p-2+\alpha}{2}} |\nabla_{\mathbb{H}}^2 u|^2 dx$$
$$+ \frac{(\alpha+1)C_p}{\varepsilon} \int_D \left(|\nabla_{\mathbb{H}} \xi|^2 + \xi|T\xi|\right) \left(\delta^2 + |\nabla_{\mathbb{H}} u|^2\right)^{\frac{p+\alpha}{2}} dx$$
$$+ \frac{(\alpha+1)^2 C_p}{\varepsilon} \int_D \xi^2 \left(\delta^2 + |\nabla_{\mathbb{H}} u|^2\right)^{\frac{p-2+\alpha}{2}} |Tu|^2 dx. \tag{5.19}$$

Now we will estimate I_1 and J_1. We have

$$X_i \left(\left(\delta^2 + |\nabla_{\mathbb{H}} u|^2\right)^{\frac{\alpha}{2}} X_1 u\right) = \frac{\alpha}{2} \left(\delta^2 + |\nabla_{\mathbb{H}} u|^2\right)^{\frac{\alpha-2}{2}} X_i \left(|\nabla_{\mathbb{H}} u|^2\right) X_1 u$$
$$+ \left(\delta^2 + |\nabla_{\mathbb{H}} u|^2\right)^{\frac{\alpha}{2}} X_i X_1 u, \tag{5.20}$$

$$X_i \left(\left(\delta^2 + |\nabla_{\mathbb{H}} u|^2\right)^{\frac{\alpha}{2}} X_2 u \right) = \frac{\alpha}{2} \left(\delta^2 + |\nabla_{\mathbb{H}} u|^2\right)^{\frac{\alpha-2}{2}} X_i \left(|\nabla_{\mathbb{H}} u|^2\right) X_2 u$$

$$+ \left(\delta^2 + |\nabla_{\mathbb{H}} u|^2\right)^{\frac{\alpha}{2}} X_i X_2 u. \tag{5.21}$$

Plugging (5.20) and (5.21) into I_1 and J_1 and using the ellipticity property (3.10) we get

$$I_1 = \int_D \xi^2 \sum_{i,j=1}^{2} \partial_{z_j} a_i(\nabla_{\mathbb{H}} u) X_j X_1 u \; X_i X_1 u \left(\delta^2 + |\nabla_{\mathbb{H}} u|^2\right)^{\frac{\alpha}{2}} \mathrm{d}x$$

$$+ \frac{\alpha}{2} \int_D \xi^2 \sum_{i,j=1}^{2} \partial_{z_j} a_i(\nabla_{\mathbb{H}} u) X_j X_1 u \; X_1 u \; X_i \left(|\nabla_{\mathbb{H}} u|^2\right) \left(\delta^2 + |\nabla_{\mathbb{H}} u|^2\right)^{\frac{\alpha-2}{2}} \mathrm{d}x$$

$$\geq \int_D \xi^2 \left(\delta^2 + |\nabla_{\mathbb{H}} u|^2\right)^{\frac{p-2+\alpha}{2}} |\nabla_{\mathbb{H}} X_1 u|^2 \mathrm{d}x$$

$$+ \frac{\alpha}{2} \int_D \xi^2 \sum_{i,j=1}^{2} \partial_{z_j} a_i(\nabla_{\mathbb{H}} u) X_j X_1 u \; X_1 u \; X_i \left(|\nabla_{\mathbb{H}} u|^2\right) \left(\delta^2 + |\nabla_{\mathbb{H}} u|^2\right)^{\frac{\alpha-2}{2}} \mathrm{d}x$$

$$\tag{5.22}$$

and

$$J_1 = \int_D \xi^2 \sum_{i,j=1}^{2} \partial_{z_j} a_i(\nabla_{\mathbb{H}} u) X_j X_2 u \; X_i X_2 u \left(\delta^2 + |\nabla_{\mathbb{H}} u|^2\right)^{\frac{\alpha}{2}} \mathrm{d}x$$

$$+ \frac{\alpha}{2} \int_D \xi^2 \sum_{i,j=1}^{2} \partial_{z_j} a_i(\nabla_{\mathbb{H}} u) X_j X_2 u \; X_2 u \; X_i \left(|\nabla_{\mathbb{H}} u|^2\right) \left(\delta^2 + |\nabla_{\mathbb{H}} u|^2\right)^{\frac{\alpha-2}{2}} \mathrm{d}x$$

$$\geq \int_D \xi^2 \left(\delta^2 + |\nabla_{\mathbb{H}} u|^2\right)^{\frac{p-2+\alpha}{2}} |\nabla_{\mathbb{H}} X_2 u|^2 \mathrm{d}x$$

$$+ \frac{\alpha}{2} \int_D \xi^2 \sum_{i,j=1}^{2} \partial_{z_j} a_i(\nabla_{\mathbb{H}} u) X_j X_2 u \; X_2 u \; X_i \left(|\nabla_{\mathbb{H}} u|^2\right) \left(\delta^2 + |\nabla_{\mathbb{H}} u|^2\right)^{\frac{\alpha-2}{2}} \mathrm{d}x.$$

$$\tag{5.23}$$

Now adding (5.22) and (5.23) and noting that $X_j X_1 u X_1 u + X_j X_2 u X_2 u = \frac{1}{2} X_j \left(|\nabla_{\mathbb{H}} u|^2\right)$ we get

$$I_1 + J_1 \geq \int_D \xi^2 \left(\delta^2 + |\nabla_{\mathbb{H}} u|^2\right)^{\frac{p-2+\alpha}{2}} |\nabla_{\mathbb{H}}^2 u|^2 \mathrm{d}x$$

$$+ \frac{\alpha}{2} \int_D \xi^2 \sum_{i,j=1}^{2} \partial_{z_j} a_i(\nabla_{\mathbb{H}} u)(X_j X_1 u \; X_1 u + X_j X_2 u \; X_2 u) X_i \left(|\nabla_{\mathbb{H}} u|^2\right)$$

$$\times \left(\delta^2 + |\nabla_{\mathbb{H}} u|^2\right)^{\frac{\alpha-2}{2}} \mathrm{d}x$$

$$\geq \int_D \xi^2 \left(\delta^2 + |\nabla_{\mathbb{H}} u|^2\right)^{\frac{p-2+\alpha}{2}} |\nabla_{\mathbb{H}}^2 u|^2 dx$$

$$+ \frac{\alpha}{4} \int_D \xi^2 \left(\delta^2 + |\nabla_{\mathbb{H}} u|^2\right)^{\frac{p-2}{2}} \left|\nabla_{\mathbb{H}}(|\nabla_{\mathbb{H}} u|^2)\right|^2 \left(\delta^2 + |\nabla_{\mathbb{H}} u|^2\right)^{\frac{\alpha-2}{2}} dx$$

$$\geq \int_D \xi^2 \left(\delta^2 + |\nabla_{\mathbb{H}} u|^2\right)^{\frac{p-2+\alpha}{2}} |\nabla_{\mathbb{H}}^2 u|^2 dx. \tag{5.24}$$

Finally adding (5.18) and (5.19), considering (5.24) we get

$$\int_D \xi^2 \left(\delta^2 + |\nabla_{\mathbb{H}} u|^2\right)^{\frac{p-2+\alpha}{2}} |\nabla_{\mathbb{H}}^2 u|^2 dx$$

$$\leq \varepsilon(\alpha+1)^2 C_p \int_D \xi^2 \left(\delta^2 + |\nabla_{\mathbb{H}} u|^2\right)^{\frac{p-2+\alpha}{2}} |\nabla_{\mathbb{H}}^2 u|^2 dx$$

$$+ \frac{(\alpha+1)C_p}{\varepsilon} \int_D \left(|\nabla_{\mathbb{H}} \xi|^2 + \xi|T\xi|\right) \left(\delta^2 + |\nabla_{\mathbb{H}} u|^2\right)^{\frac{p+\alpha}{2}} dx$$

$$+ \frac{(\alpha+1)^2 C_p}{\varepsilon} \int_D \xi^2 \left(\delta^2 + |\nabla_{\mathbb{H}} u|^2\right)^{\frac{p-2+\alpha}{2}} |Tu|^2 dx. \tag{5.25}$$

Now choosing $\varepsilon = \frac{1}{2(\alpha+1)^2 C_p}$ we can absorb the first integral of the right hand side and get the desired result. $\qquad\square$

Lemma 5.3 *Let $\alpha \geq 2$ and $\xi \in C_0^\infty(D)$. Then*

$$\int_D \xi^{\alpha+2} \left(\delta^2 + |\nabla_{\mathbb{H}} u|^2\right)^{\frac{p-2}{2}} |Tu|^\alpha |\nabla_{\mathbb{H}}^2 u|^2 dx$$

$$\leq C_p^{\frac{\alpha}{2}} (\alpha+1)^\alpha \|\nabla_{\mathbb{H}} \xi\|_{L^\infty}^\alpha \int_D \xi^2 \left(\delta^2 + |\nabla_{\mathbb{H}} u|^2\right)^{\frac{p-2+\alpha}{2}} |\nabla_{\mathbb{H}}^2 u|^2 dx. \tag{5.26}$$

Proof Following the first steps of the proof of Lemma 4.1 we have that u satisfies

$$\int_D \sum_{i=1}^2 X_1 \left(a_i(\nabla_{\mathbb{H}} u)\right) X_i \varphi \, dx - \int_D T \left(a_2(\nabla_{\mathbb{H}} u)\right) \varphi \, dx = 0 \quad \text{for all } \varphi \in HW_0^{1,2}(D). \tag{5.27}$$

Now choose $\varphi = \xi^{\alpha+2}|Tu|^\alpha X_1 u$ as a test function. Notice that this is possible because of (5.2) and (5.3). We get

$$I_1 = \int_D \sum_{i=1}^2 X_1 \left(a_i(\nabla_{\mathbb{H}} u)\right) X_i X_1 u \, \xi^{\alpha+2}|Tu|^\alpha dx$$

$$= -(\alpha+2) \int_D \sum_{i=1}^2 X_1 \left(a_i(\nabla_{\mathbb{H}} u)\right) X_i \xi \xi^{\alpha+1}|Tu|^\alpha X_1 u \, dx$$

$$- \alpha \int_D \sum_{i=1}^2 X_1 \left(a_i(\nabla_{\mathbb{H}} u) \right) X_i T u |Tu|^{\alpha-1} sign(Tu) X_1 u \xi^{\alpha+2} dx$$

$$+ \int_D T \left(a_2(\nabla_{\mathbb{H}} u) \right) \xi^{\alpha+2} |Tu|^{\alpha} X_1 u \, dx = I_2 + I_3 + II. \tag{5.28}$$

Using $X_2 X_1 = X_1 X_2 - T$ we get

$$I_1 = \int_D \sum_{i=1}^2 X_1 \left(a_i(\nabla_{\mathbb{H}} u) \right) X_1 X_i u \, \xi^{\alpha+2} |Tu|^{\alpha} dx - \int_D X_1 \left(a_2(\nabla_{\mathbb{H}} u) \right) Tu \, \xi^{\alpha+2} |Tu|^{\alpha} dx$$

$$= I_{1,1} - I_{1,2}. \tag{5.29}$$

Equation (5.28) becomes $I_{1,1} = I_{1,2} + I_2 + I_3 + II$.

$$I_{1,1} = \int_D \xi^{\alpha+2} \sum_{i,j=1}^2 \partial_{z_j} a_i(\nabla_{\mathbb{H}} u) X_1 X_j u \, X_1 X_i u |Tu|^{\alpha}$$

$$\geq c_p \int_D \xi^{\alpha+2} \left(\delta^2 + |\nabla_{\mathbb{H}} u|^2 \right)^{\frac{p-2}{2}} |\nabla_{\mathbb{H}} X_1 u|^2 |Tu|^{\alpha} dx \tag{5.30}$$

by the ellipticity property (3.10).
In $I_{1,2}$ we can integrate by parts because of the regularity result (5.2) and get

$$I_{1,2} \leq \int_D \left| a_2(\nabla_{\mathbb{H}} u) X_1 \left(\xi^{\alpha+2} Tu |Tu|^{\alpha} \right) \right| dx$$

$$\leq (\alpha + 2) C_p \int_D \xi^{\alpha+1} \left(\delta^2 + |\nabla_{\mathbb{H}} u|^2 \right)^{\frac{p-1}{2}} |X_1 \xi| |Tu|^{\alpha+1} dx$$

$$+ (\alpha + 1) C_p \int_D \xi^{\alpha+2} \left(\delta^2 + |\nabla_{\mathbb{H}} u|^2 \right)^{\frac{p-1}{2}} |Tu|^{\alpha} |X_1 Tu| \, dx$$

$$= A_1 + A_2. \tag{5.31}$$

Using Young's inequality introducing a parameter $\varepsilon > 0$ to be suitably chosen later

$$A_1 \leq (\alpha + 2) \varepsilon C_p \int_D \xi^{\alpha+2} \left(\delta^2 + |\nabla_{\mathbb{H}} u|^2 \right)^{\frac{p-2}{2}} |Tu|^{\alpha+2} dx$$

$$+ (\alpha + 2) \frac{C_p}{\varepsilon} \int_D \xi^{\alpha} |\nabla_{\mathbb{H}} \xi|^2 \left(\delta^2 + |\nabla_{\mathbb{H}} u|^2 \right)^{\frac{p}{2}} |Tu|^{\alpha} dx$$

$$\leq (\alpha + 2) \varepsilon C_p \int_D \xi^{\alpha+2} \left(\delta^2 + |\nabla_{\mathbb{H}} u|^2 \right)^{\frac{p-2}{2}} |Tu|^{\alpha} |\nabla_{\mathbb{H}}^2 u|^2 dx$$

$$+ (\alpha + 2) \frac{C_p}{\varepsilon} \int_D \xi^{\alpha} |\nabla_{\mathbb{H}} \xi|^2 \left(\delta^2 + |\nabla_{\mathbb{H}} u|^2 \right)^{\frac{p}{2}} |Tu|^{\alpha-2} |\nabla_{\mathbb{H}}^2 u|^2 dx \tag{5.32}$$

and in the last inequality we have used the fact that $|Tu| \leq 2|\nabla_{\mathbb{H}}^2 u|$. Now to estimate A_2 we use again Young's inequality, Lemma 5.1 and $|Tu| \leq 2|\nabla_{\mathbb{H}}^2 u|$:

$$A_2 \leq (\alpha+1)\frac{C_p}{\|\nabla_{\mathbb{H}}\xi\|_{L^\infty}^2}\varepsilon \int_D \xi^{\alpha+4}\left(\delta^2+|\nabla_{\mathbb{H}}u|^2\right)^{\frac{p-2}{2}}|Tu|^\alpha|\nabla_{\mathbb{H}}Tu|^2\mathrm{d}x$$

$$+ (\alpha+1)\frac{C_p}{\varepsilon}\|\nabla_{\mathbb{H}}\xi\|_{L^\infty}^2 \int_D \xi^\alpha\left(\delta^2+|\nabla_{\mathbb{H}}u|^2\right)^{\frac{p}{2}}|Tu|^\alpha\mathrm{d}x$$

$$\leq (\alpha+1)C_p\varepsilon \int_D \xi^{\alpha+2}\left(\delta^2+|\nabla_{\mathbb{H}}u|^2\right)^{\frac{p-2}{2}}|Tu|^{\alpha+2}\mathrm{d}x$$

$$+ (\alpha+1)\frac{C_p}{\varepsilon}\|\nabla_{\mathbb{H}}\xi\|_{L^\infty}^2 \int_D \xi^\alpha\left(\delta^2+|\nabla_{\mathbb{H}}u|^2\right)^{\frac{p}{2}}|Tu|^\alpha\mathrm{d}x$$

$$\leq (\alpha+1)C_p\varepsilon \int_D \xi^{\alpha+2}\left(\delta^2+|\nabla_{\mathbb{H}}u|^2\right)^{\frac{p-2}{2}}|Tu|^\alpha|\nabla_{\mathbb{H}}^2u|^2\mathrm{d}x$$

$$+ (\alpha+1)\frac{C_p}{\varepsilon}\|\nabla_{\mathbb{H}}\xi\|_{L^\infty}^2 \int_D \xi^\alpha\left(\delta^2+|\nabla_{\mathbb{H}}u|^2\right)^{\frac{p}{2}}|Tu|^{\alpha-2}|\nabla_{\mathbb{H}}^2u|^2\mathrm{d}x. \quad (5.33)$$

Now using as usual (3.11), Young's inequality and the fact that $|Tu| \leq 2|\nabla_{\mathbb{H}}^2u|$ we get

$$I_2 = (\alpha+2)\int_D \sum_{i,j=1}^2 \partial_{z_j}a_i(\nabla_{\mathbb{H}}u)X_1X_ju\ X_i\xi\ \xi^{\alpha+1}|Tu|^\alpha X_1u\ \mathrm{d}x$$

$$\leq (\alpha+2)C_p\int_D \xi^{\alpha+1}\left(\delta^2+|\nabla_{\mathbb{H}}u|^2\right)^{\frac{p-1}{2}}|\nabla_{\mathbb{H}}^2u||\nabla_{\mathbb{H}}\xi||Tu|^\alpha\mathrm{d}x$$

$$\leq (\alpha+2)C_p\varepsilon \int_D \xi^{\alpha+2}\left(\delta^2+|\nabla_{\mathbb{H}}u|^2\right)^{\frac{p-2}{2}}|Tu|^\alpha|\nabla_{\mathbb{H}}^2u|^2$$

$$+ (\alpha+2)\frac{C_p}{\varepsilon}\|\nabla_{\mathbb{H}}\xi\|_{L^\infty}^2 \int_D \xi^\alpha\left(\delta^2+|\nabla_{\mathbb{H}}u|^2\right)^{\frac{p}{2}}|Tu|^{\alpha-2}|\nabla_{\mathbb{H}}^2u|^2\mathrm{d}x. \quad (5.34)$$

Now

$$I_3 \leq \alpha\int_D \xi^{\alpha+2}\left|\sum_{i,j=1}^2 \partial_{z_j}a_i(\nabla_{\mathbb{H}}u)\right||X_1X_ju||\nabla_{\mathbb{H}}Tu||Tu|^{\alpha-1}\left(\delta^2+|\nabla_{\mathbb{H}}u|^2\right)^{\frac{1}{2}}\mathrm{d}x$$

$$\leq \alpha C_p\int_D \xi^{\alpha+2}\left(\delta^2+|\nabla_{\mathbb{H}}u|^2\right)^{\frac{p-1}{2}}|\nabla_{\mathbb{H}}^2u||\nabla_{\mathbb{H}}Tu||Tu|^{\alpha-1}\mathrm{d}x$$

$$\leq \alpha C_p\frac{\varepsilon}{\|\nabla_{\mathbb{H}}\xi\|_{L^\infty}^2} \int_D \xi^{\alpha+4}\left(\delta^2+|\nabla_{\mathbb{H}}u|^2\right)^{\frac{p-2}{2}}|Tu|^\alpha|\nabla_{\mathbb{H}}Tu|^2\mathrm{d}x$$

$$+ \frac{\alpha C_p}{\varepsilon}\|\nabla_{\mathbb{H}}\xi\|_{L^\infty}^2 \int_D \xi^\alpha\left(\delta^2+|\nabla_{\mathbb{H}}u|^2\right)^{\frac{p}{2}}|Tu|^{\alpha-2}|\nabla_{\mathbb{H}}^2u|^2\mathrm{d}x$$

$$\leq \alpha C_p\varepsilon \int_D \xi^{\alpha+2}\left(\delta^2+|\nabla_{\mathbb{H}}u|^2\right)^{\frac{p-2}{2}}|Tu|^{\alpha+2}\mathrm{d}x$$

$$+ \frac{\alpha C_p}{\varepsilon}\|\nabla_{\mathbb{H}}\xi\|_{L^\infty}^2 \int_D \xi^\alpha\left(\delta^2+|\nabla_{\mathbb{H}}u|^2\right)^{\frac{p}{2}}|Tu|^{\alpha-2}|\nabla_{\mathbb{H}}^2u|^2\mathrm{d}x$$

$$\leq \alpha C_p\varepsilon \int_D \xi^{\alpha+2}\left(\delta^2+|\nabla_{\mathbb{H}}u|^2\right)^{\frac{p-2}{2}}|Tu|^\alpha|\nabla_{\mathbb{H}}^2u|^2\mathrm{d}x$$

$$+ \frac{\alpha C_p}{\varepsilon}\|\nabla_{\mathbb{H}}\xi\|_{L^\infty}^2 \int_D \xi^\alpha\left(\delta^2+|\nabla_{\mathbb{H}}u|^2\right)^{\frac{p}{2}}|Tu|^{\alpha-2}|\nabla_{\mathbb{H}}^2u|^2\mathrm{d}x \quad (5.35)$$

where in the last two inequalities we have used Lemma 5.1 and $|Tu| \leq 2|\nabla_{\mathbb{H}}^2 u|$. Analogously to A_2

$$II = \int_D \xi^{\alpha+2} \sum_{j=1}^{2} \partial_{z_j} a_2(\nabla_{\mathbb{H}} u) T X_j u \, |Tu|^\alpha \, X_1 u \, dx$$

$$\leq C_p \int_D \xi^{\alpha+2} \left(\delta^2 + |\nabla_{\mathbb{H}} u|^2\right)^{\frac{p-1}{2}} |\nabla_{\mathbb{H}} Tu| |Tu|^\alpha dx$$

$$\leq \frac{C_p \varepsilon}{\|\nabla_{\mathbb{H}} \xi\|_{L^\infty}^2} \int_D \xi^{\alpha+4} \left(\delta^2 + |\nabla_{\mathbb{H}} u|^2\right)^{\frac{p-2}{2}} |Tu|^\alpha |\nabla_{\mathbb{H}} Tu|^2 dx$$

$$+ \frac{C_p}{\varepsilon} \|\nabla_{\mathbb{H}} \xi\|_{L^\infty}^2 \int_D \xi^\alpha \left(\delta^2 + |\nabla_{\mathbb{H}} u|^2\right)^{\frac{p}{2}} |Tu|^\alpha dx$$

$$\leq C_p \varepsilon \int_D \xi^{\alpha+2} \left(\delta^2 + |\nabla_{\mathbb{H}} u|^2\right)^{\frac{p-2}{2}} |Tu|^{\alpha+2} dx$$

$$+ \frac{C_p}{\varepsilon} \|\nabla_{\mathbb{H}} \xi\|_{L^\infty}^2 \int_D \xi^\alpha \left(\delta^2 + |\nabla_{\mathbb{H}} u|^2\right)^{\frac{p}{2}} |Tu|^\alpha dx$$

$$\leq C_p \varepsilon \int_D \xi^{\alpha+2} \left(\delta^2 + |\nabla_{\mathbb{H}} u|^2\right)^{\frac{p-2}{2}} |Tu|^\alpha |\nabla_{\mathbb{H}}^2 u|^2 dx$$

$$+ \frac{C_p}{\varepsilon} \|\nabla_{\mathbb{H}} \xi\|_{L^\infty}^2 \int_D \xi^\alpha \left(\delta^2 + |\nabla_{\mathbb{H}} u|^2\right)^{\frac{p}{2}} |Tu|^{\alpha-2} |\nabla_{\mathbb{H}}^2 u|^2 dx. \qquad (5.36)$$

Putting (5.30)–(5.36) together we get

$$\int_D \xi^{\alpha+2} \left(\delta^2 + |\nabla_{\mathbb{H}} u|^2\right)^{\frac{p-2}{2}} |X_1 \nabla_{\mathbb{H}} u|^2 |Tu|^\alpha dx$$

$$\leq (\alpha+1)\varepsilon C_p \int_D \xi^{\alpha+2} \left(\delta^2 + |\nabla_{\mathbb{H}} u|^2\right)^{\frac{p-2}{2}} |Tu|^\alpha \, |\nabla_{\mathbb{H}}^2 u|^2 dx$$

$$+ (\alpha+1) \frac{C_p}{\varepsilon} \|\nabla_{\mathbb{H}} \xi\|_{L^\infty}^2 \int_D \xi^\alpha \left(\delta^2 + |\nabla_{\mathbb{H}} u|^2\right)^{\frac{p}{2}} |Tu|^{\alpha-2} |\nabla_{\mathbb{H}}^2 u|^2 dx. \tag{5.37}$$

In an analogous way using $\varphi = \xi^{\alpha+2} |Tu|^\alpha X_2 u$ as a test function in the weak formulation of (4.53) we get

$$\int_D \xi^{\alpha+2} \left(\delta^2 + |\nabla_{\mathbb{H}} u|^2\right)^{\frac{p-2}{2}} |X_2 \nabla_{\mathbb{H}} u|^2 |Tu|^\alpha dx$$

$$\leq (\alpha+1)\varepsilon C_p \int_D \xi^{\alpha+2} \left(\delta^2 + |\nabla_{\mathbb{H}} u|^2\right)^{\frac{p-2}{2}} |Tu|^\alpha \, |\nabla_{\mathbb{H}}^2 u|^2 dx$$

$$+ (\alpha+1) \frac{C_p}{\varepsilon} \|\nabla_{\mathbb{H}} \xi\|_{L^\infty}^2 \int_D \xi^\alpha \left(\delta^2 + |\nabla_{\mathbb{H}} u|^2\right)^{\frac{p}{2}} |Tu|^{\alpha-2} |\nabla_{\mathbb{H}}^2 u|^2 dx \tag{5.38}$$

and summing (5.37) and (5.38) we get

$$
\int_D \xi^{\alpha+2} \left(\delta^2 + |\nabla_{\mathbb{H}} u|^2\right)^{\frac{p-2}{2}} |\nabla_{\mathbb{H}}^2 u|^2 |Tu|^\alpha \mathrm{d}x
$$

$$
\leq (\alpha+1)\varepsilon C_p \int_D \xi^{\alpha+2} \left(\delta^2 + |\nabla_{\mathbb{H}} u|^2\right)^{\frac{p-2}{2}} |Tu|^\alpha \, |\nabla_{\mathbb{H}}^2 u|^2 \mathrm{d}x
$$

$$
+ (\alpha+1)\frac{C_p}{\varepsilon} \|\nabla_{\mathbb{H}}\xi\|_{L^\infty}^2 \int_D \xi^\alpha \left(\delta^2 + |\nabla_{\mathbb{H}} u|^2\right)^{\frac{p}{2}} |Tu|^{\alpha-2} |\nabla_{\mathbb{H}}^2 u|^2 \mathrm{d}x.
$$

$$(5.39)$$

Now choosing $\varepsilon = \frac{1}{2(\alpha+1)C_p}$ we are able to absorb the first integral of the right hand side and get

$$
\int_D \xi^{\alpha+2} \left(\delta^2 + |\nabla_{\mathbb{H}} u|^2\right)^{\frac{p-2}{2}} |Tu|^\alpha |\nabla_{\mathbb{H}}^2 u|^2 \mathrm{d}x
$$

$$
\leq C_p(\alpha+1)^2 \|\nabla_{\mathbb{H}}\xi\|_{L^\infty}^2 \int_D \xi^\alpha \left(\delta^2 + |\nabla_{\mathbb{H}} u|^2\right)^{\frac{p}{2}} |Tu|^{\alpha-2}|\nabla_{\mathbb{H}}^2 u|^2 \mathrm{d}x.
$$

$$(5.40)$$

Using Hölder's inequality with exponent $\frac{\alpha}{\alpha-2}$ we get

$$
C_p(\alpha+1)^2 \|\nabla_{\mathbb{H}}\xi\|_{L^\infty}^2 \int_D \xi^\alpha \left(\delta^2 + |\nabla_{\mathbb{H}} u|^2\right)^{\frac{p}{2}} |Tu|^{\alpha-2}|\nabla_{\mathbb{H}}^2 u|^2 \mathrm{d}x
$$

$$
\leq C_p(\alpha+1)^2 \|\nabla_{\mathbb{H}}\xi\|_{L^\infty}^2 \left(\int_D \xi^{\alpha+2} \left(\delta^2 + |\nabla_{\mathbb{H}} u|^2\right)^{\frac{p-2}{2}} |Tu|^\alpha |\nabla_{\mathbb{H}}^2 u|^2 \mathrm{d}x\right)^{\frac{\alpha-2}{\alpha}}
$$

$$
\times \left(\int_D \xi^2 \left(\delta^2 + |\nabla_{\mathbb{H}} u|^2\right)^{\frac{p-2+\alpha}{2}} |\nabla_{\mathbb{H}}^2 u|^2 \mathrm{d}x\right)^{\frac{2}{\alpha}}.
$$

$$(5.41)$$

Now putting together (5.40) and (5.41) diving by the first factor of the right hand side of the last inequality and elevating to the power $\frac{\alpha}{2}$ we get the estimate of the theorem. $\qquad\square$

Now it's possible to get rid of the second integral of the right hand side of Lemma 5.2 and get an estimate involving only horizontal derivatives. This will allow us to use the subelliptic version of Sobolev's embedding theorem and eventually run Moser's iteration.

Lemma 5.4 *Let $\alpha \geq 2$ and $\xi \in C_0^\infty(D)$. Then*

$$
\int_D \xi^2 \left(\delta^2 + |\nabla_{\mathbb{H}} u|^2\right)^{\frac{p-2+\alpha}{2}} |\nabla_{\mathbb{H}}^2 u|^2 \mathrm{d}x \leq C_p K_\xi (\alpha+1)^{10} \int_{\mathrm{supp}\xi} \left(\delta^2 + |\nabla_{\mathbb{H}} u|^2\right)^{\frac{p+\alpha}{2}} \mathrm{d}x
$$

where $K_\xi = \|\nabla_{\mathbb{H}}\xi\|_{L^\infty}^2 + \xi \|T\xi\|_{L^\infty}$.

Proof We just need to estimate the second integral in Lemma 5.2. We use Hölder's inequality with exponent $\frac{\alpha+2}{2}$ to obtain a term equal to the one on the left hand side of Lemma 5.3:

$$\int_D \xi^2 \left(\delta^2 + |\nabla_{\mathbb{H}} u|^2\right)^{\frac{p-2+\alpha}{2}} |Tu|^2 dx$$

$$\leq \left(\int_D \xi^{\alpha+2} \left(\delta^2 + |\nabla_{\mathbb{H}} u|^2\right)^{\frac{p-2}{2}} |Tu|^{\alpha+2} dx\right)^{\frac{2}{\alpha+2}}$$

$$\times \left(\int_{\text{supp}\xi} \left(\delta^2 + |\nabla_{\mathbb{H}} u|^2\right)^{\frac{p+\alpha}{2}} dx\right)^{\frac{\alpha}{\alpha+2}}$$

$$\leq \left(\int_D \xi^{\alpha+2} \left(\delta^2 + |\nabla_{\mathbb{H}} u|^2\right)^{\frac{p-2}{2}} |Tu|^{\alpha} |\nabla_{\mathbb{H}}^2 u|^2 dx\right)^{\frac{2}{\alpha+2}}$$

$$\times \left(\int_{\text{supp}\xi} \left(\delta^2 + |\nabla_{\mathbb{H}} u|^2\right)^{\frac{p+\alpha}{2}} dx\right)^{\frac{\alpha}{\alpha+2}}$$

$$\leq C_p^{\frac{\alpha}{\alpha+2}} (\alpha+1)^{\frac{2\alpha}{\alpha+2}} \|\nabla_{\mathbb{H}}\xi\|_{L^\infty(D)}^{\frac{2\alpha}{\alpha+2}} \left(\int_D \xi^2 \left(\delta^2 + |\nabla_{\mathbb{H}} u|^2\right)^{\frac{p-2+\alpha}{2}} |\nabla_{\mathbb{H}}^2 u|^2 dx\right)^{\frac{2}{\alpha+2}}$$

$$\times \left(\int_{\text{supp}\xi} \left(\delta^2 + |\nabla_{\mathbb{H}} u|^2\right)^{\frac{p+\alpha}{2}} dx\right)^{\frac{\alpha}{\alpha+2}} dx \qquad (5.42)$$

where in the last inequality we have used Lemma 5.3. Now let

$$A = \int_D \xi^2 \left(\delta^2 + |\nabla_{\mathbb{H}} u|^2\right)^{\frac{p-2+\alpha}{2}} |\nabla_{\mathbb{H}}^2 u|^2 dx \qquad (5.43)$$

$$B = \int_{\text{supp}\xi} \left(\delta^2 + |\nabla_{\mathbb{H}} u|^2\right)^{\frac{p+\alpha}{2}} dx \qquad (5.44)$$

and $p = \frac{\alpha+2}{2}$, $q = \frac{\alpha+2}{\alpha}$. With these notations Lemma 5.2, (5.42) and Young's inequality imply

$$A \leq c_1 B + (\alpha+1)^4 c_2 A^{\frac{1}{p}} B^{\frac{1}{q}} \leq c_1 B + \varepsilon(\alpha+1)^4 \frac{A}{p} + (\alpha+1)^4 \frac{c_2^q}{\varepsilon^{\frac{q}{p}} q} B$$

where $c_1 = C_p(\alpha+1)^3 \left\| |\nabla_{\mathbb{H}}\xi|^2 + \xi|T\xi| \right\|_{L^\infty(D)}$ and $c_2 = C^{\frac{\alpha}{\alpha+2}} (\alpha+1)^{\frac{2\alpha}{\alpha+2}} \|\nabla_{\mathbb{H}}\xi\|_{L^\infty(D)}^{\frac{2\alpha}{\alpha+2}}$. Now choosing $\varepsilon = \frac{p}{2(\alpha+1)^4}$ we can bring $\varepsilon(\alpha+1)^4 \frac{A}{p}$ to left hand side and get the result, after noting that $6 + \frac{8}{\alpha} \leq 10$ for $\alpha \geq 2$. $\qquad \square$

5.2 Lipschitz Estimate

In this section we are going to prove Lipschitz estimates for solutions of the non degenerate p-Laplace equation (3.1) which are independent of the non degeneracy parameter, so that we can obtain the same estimates also in the degenerate case. We first work in a domain D satisfying (3.34) so that we can apply all the results of the previous section, and then we will proceed with the general case. The main tools, as in the Euclidean case, will be Sobolev's embedding theorem and the preceding estimates which will allow us to run an adapted version of Moser's iteration.

Theorem 5.1 *Let $u \in HW^{1,p}(D)$, $1 < p < \infty$ be a weak solution of the non degenerate p-Laplace equation (3.1). Then*

$$\|\nabla_{\mathbb{H}} u\|_{L^{\infty}(B_r)} \le C_p \left(\fint_{B_{2r}} \left(\delta^2 + |\nabla_{\mathbb{H}} u|^2 \right)^{\frac{p}{2}} \mathrm{d}x \right)^{\frac{1}{p}} \tag{5.45}$$

for every ball B_r such that the concentric ball $B_{2r} \subset D$.

Proof Observe that

$$\int_D \left| \nabla_{\mathbb{H}} \left(\xi \left(\delta^2 + |\nabla_{\mathbb{H}} u|^2 \right)^{\frac{p+\alpha}{4}} \right) \right|^2 \mathrm{d}x$$

$$= \int_D |\nabla_{\mathbb{H}} \xi|^2 \left(\delta^2 + |\nabla_{\mathbb{H}} u|^2 \right)^{\frac{p+\alpha}{2}} \mathrm{d}x$$

$$+ \left(\frac{p+\alpha}{4} \right)^2 \int_D \xi^2 \left(\delta^2 + |\nabla_{\mathbb{H}} u|^2 \right)^{\frac{p-4+\alpha}{2}} \left| \nabla_{\mathbb{H}} \left(\delta^2 + |\nabla_{\mathbb{H}} u|^2 \right) \right|^2 \mathrm{d}x$$

$$= I_1 + I_2. \tag{5.46}$$

We have

$$\frac{1}{4} \left| \nabla_{\mathbb{H}} \left(\delta^2 + |\nabla_{\mathbb{H}} u|^2 \right) \right|^2$$

$$= |X_1 X_1 u X_1 u + X_1 X_2 u X_2 u|^2 + |X_2 X_1 u X_1 u + X_2 X_2 u X_2 u|^2$$

$$\le \left(|X_1 X_1 u|^2 + |X_2 X_1 u|^2 \right) |X_1 u|^2 + \left(|X_1 X_2 u|^2 + |X_2 X_2 u|^2 \right) |X_2 u|^2$$

$$\le \left(|X_1 X_1 u|^2 + |X_2 X_1 u|^2 + |X_1 X_2 u|^2 + |X_2 X_2 u|^2 \right) |\nabla_{\mathbb{H}} u|^2$$

$$\le |\nabla_{\mathbb{H}}^2 u|^2 |\nabla_{\mathbb{H}} u|^2. \tag{5.47}$$

Now using Lemma 5.4 we obtain

$$I_2 \le (p+\alpha)^2 \int_D \xi^2 \left(\delta^2 + |\nabla_{\mathbb{H}} u|^2 \right)^{\frac{p-2+\alpha}{2}} |\nabla_{\mathbb{H}}^2 u|^2 \mathrm{d}x$$

$$\le C_p K_\xi (p+\alpha)^{12} \int_{\mathrm{supp}\xi} \left(\delta^2 + |\nabla_{\mathbb{H}} u|^2 \right)^{\frac{p+\alpha}{2}} \mathrm{d}x. \tag{5.48}$$

From (5.46) and (5.48) we get

$$\int_D \left| \nabla_{\mathbb{H}} \left(\xi \left(\delta^2 + |\nabla_{\mathbb{H}} u|^2 \right)^{\frac{p+\alpha}{4}} \right) \right|^2 dx \leq C_p K_\xi (p + \alpha)^{12} \int_{\text{supp}\xi} \left(\delta^2 + |\nabla_{\mathbb{H}} u|^2 \right)^{\frac{p+\alpha}{2}} dx$$

(5.49)

where K_ξ is the constant appearing in Lemma 5.4. Fix two concentric balls $B_{\lambda r} \subset B_r \subset D, 0 < \lambda < 1$ and consider a sequence of decreasing radii

$$r_i = \lambda r + \frac{r - \lambda r}{2^i} \searrow \lambda r$$

and cut-off functions $\xi_i \in C_0^\infty (B_{r_i})$ such that $\xi_i \equiv 1$ in $B_{r_{i+1}}$ and $\|\nabla \xi_i\|_{L^\infty} \leq \frac{c}{r_i - r_{i+1}}$. Now using the subelliptic version of Sobolev's embedding Theorem 2.8 we get

$$\left(\fint_{B_{r_i}} \left(\xi \left(\delta^2 + |\nabla_{\mathbb{H}} u|^2 \right)^{\frac{p+\alpha}{4}} \right)^{\frac{2Q}{Q-2}} dx \right)^{\frac{Q-2}{Q}} \leq C r_i^2 \fint_{B_{r_i}} \left| \nabla_{\mathbb{H}} \left(\xi \left(\delta^2 + |\nabla_{\mathbb{H}} u|^2 \right)^{\frac{p+\alpha}{4}} \right) \right|^2 dx.$$

(5.50)

Putting together (5.49) and (5.50), calling $k = \frac{Q}{Q-2}$ and using the properties of the cut-off functions ξ_i we have

$$\left(\fint_{B_{r_{i+1}}} \left(\delta^2 + |\nabla_{\mathbb{H}} u|^2 \right)^{\frac{p+\alpha}{2} k} dx \right)^{\frac{1}{k}} \leq C_p \left(\frac{r_i}{r_i - r_{i+1}} \right)^2 (p + \alpha)^{12}$$

$$\times \fint_{B_{r_i}} \left(\delta^2 + |\nabla_{\mathbb{H}} u|^2 \right)^{\frac{p+\alpha}{2}} dx$$

$$\leq \frac{C_p}{(1 - \lambda)^2} (p + \alpha)^{12} \fint_{B_{r_i}} \left(\delta^2 + |\nabla_{\mathbb{H}} u|^2 \right)^{\frac{p+\alpha}{2}} dx.$$

(5.51)

Consider a sequence of increasing exponents

$$\alpha_i = (p + 2)k^i - p \geq 2$$

so we can use it in (5.51) to get

$$\left(\fint_{B_{r_{i+1}}} \left(\delta^2 + |\nabla_{\mathbb{H}} u|^2 \right)^{\frac{p+2}{2} k^{i+1}} dx \right)^{\frac{1}{(p+2)k^{i+1}}}$$

$$\leq \left(\frac{C_p}{(1 - \lambda)^2} \right)^{\frac{1}{(p+2)k^i}} \left((p + 2)k^i \right)^{\frac{12}{(p+2)k^i}} \left(\fint_{B_{r_i}} \left(\delta^2 + |\nabla_{\mathbb{H}} u|^2 \right)^{\frac{p+2}{2} k^i} dx \right)^{\frac{1}{(p+2)k^i}}$$

(5.52)

where we have also raised both sides to the power $\frac{1}{(p+2)k^i}$. Now to simplify the notation call $\beta_i = (p+2)k^i$ and iterating (5.52) we get by induction

$$
\left(\fint_{B_{r_{n+1}}} \left(\delta^2 + |\nabla_{\mathbb{H}} u|^2 \right)^{\frac{\beta_{n+1}}{2}} dx \right)^{\frac{1}{\beta_{n+1}}} \leq \left(\frac{C_p}{(1-\lambda)^2} \right)^{\sum_{i=0}^{n} \frac{1}{\beta_i}}
$$
$$
\times \prod_{i=0}^{n} \beta_i^{\frac{12}{\beta_i}} \left(\fint_{B_r} \left(\delta^2 + |\nabla_{\mathbb{H}} u|^2 \right)^{\frac{p+2}{2}} dx \right)^{\frac{1}{p+2}}
$$
$$
\leq \left(\frac{C_p}{(1-\lambda)^2} \right)^{\sum_{i=0}^{\infty} \frac{1}{\beta_i}}
$$
$$
\times \prod_{i=0}^{\infty} \beta_i^{\frac{12}{\beta_i}} \left(\fint_{B_r} \left(\delta^2 + |\nabla_{\mathbb{H}} u|^2 \right)^{\frac{p+2}{2}} dx \right)^{\frac{1}{p+2}}.
$$

(5.53)

Now

$$
\sum_{i=0}^{\infty} \frac{1}{\beta_i} = \frac{1}{p+2} \sum_{i=0}^{\infty} \frac{1}{k^i} = \frac{k}{(k-1)(p+2)} = \frac{Q}{2(p+2)} \qquad (5.54)
$$

and

$$
\log \left(\prod_{i=0}^{\infty} \beta_i^{\frac{12}{\beta_i}} \right) = \sum_{i=0}^{\infty} \frac{12}{\beta_i} \log \beta_i = 12 \frac{\log(p+2)}{p+2} \frac{k}{k-1} + 12 \log k \sum_{i=0}^{\infty} \frac{i}{k^i} \quad (5.55)
$$

which is a finite constant depending only on p and Q. In the end we have

$$
\left(\fint_{B_{r_{n+1}}} \left(\delta^2 + |\nabla_{\mathbb{H}} u|^2 \right)^{\frac{\beta_{n+1}}{2}} dx \right)^{\frac{1}{\beta_{n+1}}} \leq \frac{C_p}{(1-\lambda)^{\frac{Q}{p+2}}} \left(\fint_{B_r} \left(\delta^2 + |\nabla_{\mathbb{H}} u|^2 \right)^{\frac{p+2}{2}} dx \right)^{\frac{1}{p+2}}.
$$

(5.56)

Now since β_n tends to infinity when n tends to infinity, and since the averages on the left hand side of the previous inequality tend to the essential supremum of the integrand we get

$$
\|\nabla_{\mathbb{H}} u\|_{L^\infty(B_{\lambda r})} \leq \frac{C_p}{(1-\lambda)^{\frac{Q}{p+2}}} \left(\fint_{B_r} \left(\delta^2 + |\nabla_{\mathbb{H}} u|^2 \right)^{\frac{p+2}{2}} dx \right)^{\frac{1}{p+2}} \qquad (5.57)
$$

and this is valid for all $B_r \subset D$ and for all $0 < \lambda < 1$.

Since $\left\| (\delta^2 + |\nabla_{\mathbb{H}} u|^2) \right\|_{L^s(B_{\lambda r})} \nearrow \left\| (\delta^2 + |\nabla_{\mathbb{H}} u|^2) \right\|_{L^\infty(B_{\lambda r})}$ as $s \to \infty$ we have

$$\left(\fint_{B_{\lambda r}} (\delta^2 + |\nabla_{\mathbb{H}} u|^2)^{\frac{s}{2}} \, dx \right)^{\frac{1}{s}} \le \frac{C_p}{(1 - \lambda)^{\frac{Q}{p+2}}} \left(\fint_{B_r} (\delta^2 + |\nabla_{\mathbb{H}} u|^2)^{\frac{p+2}{2}} \, dx \right)^{\frac{1}{p+2}} \quad (5.58)$$

for every $s > p + 2$ and $0 < \lambda < 1$. Now we want to show that

$$\left(\fint_{B_{\lambda r}} (\delta^2 + |\nabla_{\mathbb{H}} u|^2)^{\frac{s}{2}} \, dx \right)^{\frac{1}{s}} \le \frac{C_{p,\theta}}{(1 - \lambda)^{\frac{Q}{(p+2)\theta}}} \left(\fint_{B_r} (\delta^2 + |\nabla_{\mathbb{H}} u|^2)^{\frac{p}{2}} \, dx \right)^{\frac{1}{p}} \quad (5.59)$$

with $\frac{1}{p+2} = \frac{\theta}{p} + \frac{1-\theta}{s}$. Let $\zeta = \frac{Q}{p+2}$ and

$$\Phi = \sup_{\frac{1}{2} < \lambda < 1} (1 - \lambda)^{\zeta \frac{1-\theta}{\theta}} \left(\fint_{B_{\lambda r}} (\delta^2 + |\nabla_{\mathbb{H}} u|^2)^{\frac{p+2}{2}} \, dx \right)^{\frac{1}{p+2}}. \quad (5.60)$$

Now choose $\frac{1}{2} < \lambda' := \frac{\lambda+1}{2} < 1$, so that we have

$$\frac{(1 - \lambda)^{\frac{\zeta}{\theta}}}{\left(1 - \frac{\lambda}{\lambda'}\right)^\zeta} \le 2^{\frac{\zeta}{\theta}} (1 - \lambda')^{\frac{\zeta}{\theta}(1-\theta)}. \quad (5.61)$$

Indeed

$$\frac{(1 - \lambda)^{\frac{\zeta}{\theta}}}{\left(1 - \frac{\lambda}{\lambda'}\right)^\zeta (1 - \lambda')^{\zeta \frac{1-\theta}{\theta}}} = (\lambda + 1)^\zeta 2^{\frac{\zeta}{\theta}(1-\theta)} \le 2^{\frac{\zeta}{\theta}}. \quad (5.62)$$

Now from (5.61), (5.58) and the definition of Φ we have

$$(1 - \lambda)^{\frac{\zeta}{\theta}} \left(\fint_{B_{\lambda r}} (\delta^2 + |\nabla_{\mathbb{H}} u|^2)^{\frac{s}{2}} \, dx \right)^{\frac{1}{s}}$$

$$\le C_p \frac{(1 - \lambda)^{\frac{\zeta}{\theta}}}{\left(1 - \frac{\lambda}{\lambda'}\right)^\zeta} \left(\fint_{B_{\lambda' r}} (\delta^2 + |\nabla_{\mathbb{H}} u|^2)^{\frac{p+2}{2}} \, dx \right)^{\frac{1}{p+2}}$$

$$\le C_p 2^{\frac{\zeta}{\theta}} (1 - \lambda')^{\frac{\zeta}{\theta}(1-\theta)} \left(\fint_{B_{\lambda' r}} (\delta^2 + |\nabla_{\mathbb{H}} u|^2)^{\frac{p+2}{2}} \, dx \right)^{\frac{1}{p+2}}$$

$$\le C_p 2^{\frac{\zeta}{\theta}} \Phi. \quad (5.63)$$

Take $\eta > 0$, then by the definition of Φ there exists $\lambda' \in]\frac{1}{2}, 1[$ such that

$$(1 - \lambda')^{\frac{\zeta}{\theta}(1-\theta)} \left(\fint_{B_{\lambda' r}} (\delta^2 + |\nabla_{\mathbb{H}} u|^2)^{\frac{p+2}{2}} \, dx \right)^{\frac{1}{p+2}} \ge \Phi - \eta$$

so

$$\Phi \leq (1-\lambda')^{\frac{\zeta}{\theta}(1-\theta)} \left(\fint_{B_{\lambda'r}} \left(\delta^2 + |\nabla_{\mathbb{H}} u|^2\right)^{\frac{p+2}{2}} dx \right)^{\frac{1}{p+2}} + \eta$$

$$= (1-\lambda')^{\frac{\zeta}{\theta}(1-\theta)} \left(\fint_{B_{\lambda'r}} \left(\delta^2 + |\nabla_{\mathbb{H}} u|^2\right)^{\frac{p+2}{2}\theta} \left(\delta^2 + |\nabla_{\mathbb{H}} u|^2\right)^{\frac{p+2}{2}(1-\theta)} dx \right)^{\frac{1}{p+2}} + \eta$$

$$\leq (1-\lambda')^{\frac{\zeta}{\theta}(1-\theta)} \left(\fint_{B_{\lambda'r}} \left(\delta^2 + |\nabla_{\mathbb{H}} u|^2\right)^{\frac{p}{2}} dx \right)^{\frac{\theta}{p}} \left(\fint_{B_{\lambda'r}} \left(\delta^2 + |\nabla_{\mathbb{H}} u|^2\right)^{\frac{s}{2}} dx \right)^{\frac{1-\theta}{s}} + \eta$$

$$(5.64)$$

where we have applied Hölder's inequality with exponents $\frac{p}{\theta(p+2)}$ and $\frac{s}{(1-\theta)(p+2)}$ since $1 = \theta\frac{p+2}{p} + (1-\theta)\frac{p+2}{s}$. Now applying Young's inequality with exponents $\frac{1}{\theta}$ and $\frac{1}{1-\theta}$ introducing also a parameter $\varepsilon > 0$ we get

$$(1-\lambda')^{\frac{\zeta}{\theta}(1-\theta)} \left(\fint_{B_{\lambda'r}} \left(\delta^2 + |\nabla_{\mathbb{H}} u|^2\right)^{\frac{p}{2}} dx \right)^{\frac{\theta}{p}} \left(\fint_{B_{\lambda'r}} \left(\delta^2 + |\nabla_{\mathbb{H}} u|^2\right)^{\frac{s}{2}} dx \right)^{\frac{1-\theta}{s}}$$

$$\leq \frac{\theta}{\varepsilon^{\frac{1-\theta}{\theta}}} \left(\fint_{B_{\lambda'r}} \left(\delta^2 + |\nabla_{\mathbb{H}} u|^2\right)^{\frac{p}{2}} dx \right)^{\frac{1}{p}} + \varepsilon(1-\theta)(1-\lambda')^{\frac{\zeta}{\theta}} \left(\fint_{B_{\lambda'r}} \left(\delta^2 + |\nabla_{\mathbb{H}} u|^2\right)^{\frac{s}{2}} dx \right)^{\frac{1}{s}}$$

$$\leq \frac{2^Q \theta}{\varepsilon^{\frac{1-\theta}{\theta}}} \left(\fint_{B_r} \left(\delta^2 + |\nabla_{\mathbb{H}} u|^2\right)^{\frac{p}{2}} dx \right)^{\frac{1}{p}} + \varepsilon(1-\theta)\Phi$$

$$(5.65)$$

where we have used $\frac{|B_r|}{|B_{\lambda'r}|} = \frac{1}{\lambda'^Q} \leq 2^Q$. Now putting together (5.64) and (5.65) and choosing $\varepsilon = \frac{1}{2(1-\theta)}$ we get

$$\Phi \leq C_Q \theta(1-\theta)^{\frac{1-\theta}{\theta}} \left(\fint_{B_r} \left(\delta^2 + |\nabla_{\mathbb{H}} u|^2\right)^{\frac{p}{2}} dx \right)^{\frac{1}{p}}.$$

Finally considering (5.63) we get

$$(1-\lambda)^{\frac{\zeta}{\theta}} \left(\fint_{B_{\lambda r}} \left(\delta^2 + |\nabla_{\mathbb{H}} u|^2\right)^{\frac{s}{2}} dx \right)^{\frac{1}{2}} \leq C_{p,Q} \theta(1-\theta)^{\frac{1-\theta}{\theta}} 2^{\frac{\zeta}{\theta}} \left(\fint_{B_r} \left(\delta^2 + |\nabla_{\mathbb{H}} u|^2\right)^{\frac{p}{2}} dx \right)^{\frac{1}{p}}$$

which, passing to the limit for $s \to \infty$ gives

$$\|\nabla_{\mathbb{H}} u\|_{L^\infty(B_{\lambda r})} \leq \frac{C_{p,Q}}{(1-\lambda)^{\frac{Q}{p}}} \left(\fint_{B_r} \left(\delta^2 + |\nabla_{\mathbb{H}} u|^2\right)^{\frac{p}{2}} \right)^{\frac{1}{p}} \quad (5.66)$$

since $\theta(s) \to \frac{p}{p+2}$ when $s \to \infty$. $\qquad \square$

Now we consider a general open subset Ω of the Heisenberg group $\mathbb{H}$. We will be using an approximation argument, producing solutions which satisfy the hypothesis of the previous theorem in suitable balls, therefore we will have a local result.

Theorem 5.2 (Lipschitz estimate, non degenerate case) Let $u \in HW^{1,p}(\Omega)$, $1 < p < \infty$ be a weak solution of the non degenerate p-Laplace equation (3.1). Then

$$\|\nabla_{\mathbb{H}}u\|_{L^\infty(B_r)} \le C_p \left(\fint_{B_{2r}} \left(\delta^2 + |\nabla_{\mathbb{H}}u|^2\right)^{\frac{p}{2}} dx \right)^{\frac{1}{p}} \tag{5.67}$$

for every ball B_r such that the concentric ball $B_{2r} \subset \Omega$.

Proof Fix a Carnot-Carathéodory ball $B_r \subset \Omega$. Since $C^\infty(B_r) \cap HW^{1,p}(B_r)$ is dense in $HW^{1,p}(B_r)$ there exists a sequence of regular functions $\psi_\varepsilon \in C^\infty(B_r) \cap HW^{1,p}(B_r)$ that converges to u in $HW^{1,p}(B_r)$. Take an Euclidean ball $B^E_{\sigma r} \subset B_r$: it satisfies condition (3.34). Consider the Dirichlet problems

$$\begin{cases} \operatorname{div}_{\mathbb{H}}\left(\left(\delta^2 + |\nabla_{\mathbb{H}}v|^2\right)^{\frac{p-2}{2}} \nabla_{\mathbb{H}}v\right) = 0 & \text{in } B^E_{\sigma r} \\ v - \psi_\varepsilon \in HW^{1,p}_0(B^E_{\sigma r}) \end{cases} \tag{5.68}$$

and denote u_ε the solution of those problems (note that there exists a solution and it is unique by virtue of Theorem 3.3). Next use $\varphi_\varepsilon = u_\varepsilon - \psi_\varepsilon$ as a test function in the weak formulation of (5.68) to get

$$\int_{B^E_{\sigma r}} \left(\delta^2 + |\nabla_{\mathbb{H}}u_\varepsilon|^2\right)^{\frac{p-2}{2}} |\nabla_{\mathbb{H}}u_\varepsilon|^2 \, dx$$

$$= \int_{B^E_{\sigma r}} \left(\delta^2 + |\nabla_{\mathbb{H}}u_\varepsilon|^2\right)^{\frac{p-2}{2}} \langle \nabla_{\mathbb{H}}u_\varepsilon, \nabla_{\mathbb{H}}\psi_\varepsilon \rangle \, dx$$

$$\le \int_{B^E_{\sigma r}} \left(\delta^2 + |\nabla_{\mathbb{H}}u_\varepsilon|^2\right)^{\frac{p-1}{2}} |\nabla_{\mathbb{H}}\psi_\varepsilon| \, dx. \tag{5.69}$$

Now

$$\int_{B^E_{\sigma r}} \left(\delta^2 + |\nabla_{\mathbb{H}}u_\varepsilon|^2\right)^{\frac{p}{2}} dx = \int_{B^E_{\sigma r}} \left(\delta^2 + |\nabla_{\mathbb{H}}u_\varepsilon|^2\right)^{\frac{p-2}{2}} \left(\delta^2 + |\nabla_{\mathbb{H}}u_\varepsilon|^2\right) dx$$

$$= \int_{B^E_{\sigma r}} \left(\delta^2 + |\nabla_{\mathbb{H}}u_\varepsilon|^2\right)^{\frac{p-2}{2}} \delta^2 dx$$

$$+ \int_{B^E_{\sigma r}} \left(\delta^2 + |\nabla_{\mathbb{H}}u_\varepsilon|^2\right)^{\frac{p-2}{2}} |\nabla_{\mathbb{H}}u_\varepsilon|^2$$

$$\le \eta C_p \int_{B^E_{\sigma r}} \left(\delta^2 + |\nabla_{\mathbb{H}}u_\varepsilon|^2\right)^{\frac{p}{2}} dx$$

$$+ \frac{C_p}{\eta^{\frac{p-2}{2}}} \int_{B^E_{\sigma r}} \delta^p dx + \int_{B^E_{\sigma r}} \left(\delta^2 + |\nabla_{\mathbb{H}}u_\varepsilon|^2\right)^{\frac{p-1}{2}} |\nabla_{\mathbb{H}}\psi_\varepsilon| dx$$

$$\leq \eta C_p \int_{B_{\sigma r}^E} \left(\delta^2 + |\nabla_{\mathbb{H}} u_\varepsilon|^2\right)^{\frac{p}{2}} dx + \frac{C_p}{\eta^{\frac{p-2}{2}}} \int_{B_{\sigma r}^E} \delta^p dx$$

$$+ \eta C_p \int_{B_{\sigma r}^E} \left(\delta^2 + |\nabla_{\mathbb{H}} u_\varepsilon|^2\right)^{\frac{p}{2}} dx + \frac{C_p}{\eta^{p-1}} \int_{B_{\sigma r}^E} |\nabla_{\mathbb{H}} \psi_\varepsilon|^p dx$$

$$(5.70)$$

where we have used (5.69) and Young's inequality twice with exponents respectively $\frac{p}{p-2}$ and $\frac{p}{p-1}$ and a parameter $\eta > 0$ that we now choose to be $\eta = \frac{1}{2C_p}$. This gives

$$\int_{B_{\sigma r}^E} \left(\delta^2 + |\nabla_{\mathbb{H}} u_\varepsilon|^2\right)^{\frac{p}{2}} dx \leq C_p \int_{B_{\sigma r}^E} \delta^p dx + C_p \int_{B_{\sigma r}^E} \left(\delta^2 + |\nabla_{\mathbb{H}} \psi_\varepsilon|^2\right)^{\frac{p}{2}} dx$$

$$\leq C_p \int_{B_{\sigma r}^E} \left(\delta^2 + |\nabla_{\mathbb{H}} \psi_\varepsilon|^2\right)^{\frac{p}{2}} dx. \qquad (5.71)$$

Moreover

$$\int_{B_{\sigma r}^E} \left(\delta^2 + |\nabla_{\mathbb{H}} \psi_\varepsilon|^2\right)^{\frac{p}{2}} dx \leq \int_{B_{\sigma r}^E} \left(\delta^2 + |\nabla_{\mathbb{H}} u|^2\right)^{\frac{p}{2}} + \int_{B_{\sigma r}^E} |\nabla_{\mathbb{H}} u - \nabla_{\mathbb{H}} \psi_\varepsilon|^p dx. \quad (5.72)$$

From (5.71) and (5.72) we get

$$\int_{B_{\sigma r}^E} \left(\delta^2 + |\nabla_{\mathbb{H}} u_\varepsilon|^2\right)^{\frac{p}{2}} dx \leq \int_{B_{\sigma r}^E} \left(\delta^2 + |\nabla_{\mathbb{H}} u|^2\right)^{\frac{p}{2}} + 1 \qquad (5.73)$$

for ε sufficiently small because $\psi_\varepsilon \to u$ in $HW^{1,p}(B_r)$. This implies that $\|u_\varepsilon\|_{HW^{1,p}(B_{\sigma r}^E)}$ is uniformly bounded with respect to ε and since $1 < p < \infty$ we can extract a weakly convergent subsequence (which we continue to denote by u_ε) in $HW^{1,p}(B_{\sigma r}^E)$. Namely there exists $\bar{u} \in HW^{1,p}(B_{\sigma r}^E)$ such that $u_\varepsilon \xrightarrow{w} \bar{u}$ in $HW^{1,p}(B_{\sigma r}^E)$. Since $u_\varepsilon - \psi_\varepsilon \in HW_0^{1,p}(B_{\sigma r}^E)$ and this space is closed under weak convergence we get that $\bar{u} - u \in HW_0^{1,p}(B_{\sigma r}^E)$. Now from Lemma 3.6 we get that $\nabla_{\mathbb{H}} u_\varepsilon \to \nabla_{\mathbb{H}} \bar{u}$ in $L^p(B_{\sigma r}^E)$ and $\bar{u}$ is a weak solution of the p-Laplace equation (3.1) in $B_{\sigma r}^E$. Since the solution of this equation among functions who take the same boundary value in the Sobolev sense is unique (Theorem 3.3) and since $\bar{u} - u \in HW_0^{1,p}(B_{\sigma r})$ we get that $\bar{u} = u$. Now applying Theorem 3.5 we can conclude that

$$\|\nabla_{\mathbb{H}} u_\varepsilon\|_{L^\infty(B_{\sigma r}^E)} \leq M$$

with a positive constant M which has the dependences described in Remark 3.2. Under this additional property we can use Theorem 5.1 to get

$$\|\nabla_{\mathbb{H}} u_\varepsilon\|_{L^\infty(B_{\tau r})} \leq C_p \left(\fint_{B_{2\tau r}} \left(\delta^2 + |\nabla_{\mathbb{H}} u_\varepsilon|^2\right)^{\frac{p}{2}} dx\right)^{\frac{1}{p}} \qquad (5.74)$$

for any Carnot-Carathéodory ball $B_{\tau r}$ such that the concentric ball $B_{2\tau r} \subset B_{\sigma r}^E$. Since $\nabla_{\mathbb{H}} u_\varepsilon \to \nabla_{\mathbb{H}} u$ pointwisely a.e. in $B_{\sigma r}^E$ we can conclude from (5.74) and (5.72) that

$$\|\nabla_{\mathbb{H}} u\|_{L^\infty(B_{\tau r})} \leq C_p \left(\fint_{B_r} \left(\delta^2 + |\nabla_{\mathbb{H}} u|^2 \right)^{\frac{p}{2}} dx \right)^{\frac{1}{p}}. \tag{5.75}$$

Using a covering argument we get the result.   □

Theorem 5.3 (Lipschitz estimate, degenerate case) *Let* $u \in HW^{1,p}(\Omega)$, $1 < p < \infty$ *be a weak solution of the degenerate p-Laplace equation (3.1). Then*

$$\|\nabla_{\mathbb{H}} u\|_{L^\infty(B_r)} \leq C_p \left(\fint_{B_{2r}} |\nabla_{\mathbb{H}} u|^p dx \right)^{\frac{1}{p}} \tag{5.76}$$

for every ball B_r such that the concentric ball $B_{2r} \subset \Omega$.

Proof Let u_δ be the unique weak solution of the Dirichlet problem

$$\begin{cases} \operatorname{div}_{\mathbb{H}} \left(\left(\delta^2 + |\nabla_{\mathbb{H}} u_\delta| \right)^{\frac{p-2}{2}} \nabla_{\mathbb{H}} u_\delta \right) = 0 & \text{in } \Omega \\ u_\delta - u \in HW_0^{1,p}(\Omega) \end{cases} \tag{5.77}$$

where $u \in HW_0^{1,p}(\Omega)$ is the solution of the (degenerate) p-Laplace equation (3.1) with some boundary condition $g \in HW^{1,p}(\Omega)$. By Theorem 3.2 u_δ is the minimum of the Dirichlet functional $\mathcal{D}_{p,\delta}$ in $\mathcal{A}_u = \left\{ v \in HW^{1,p}(\Omega) \mid v - u \in HW_0^{1,p}(\Omega) \right\}$ and u is the minimum of the Dirichlet functional $\mathcal{D}_p = \mathcal{D}_{p,0}$ in $\mathcal{A}_g$. Therefore we have

$$\mathcal{D}_p(u) \leq \mathcal{D}_p(u_\delta) \leq \mathcal{D}_{p,\delta}(u_\delta) \leq \mathcal{D}_{p,\delta}(u) \tag{5.78}$$

and subtracting $\mathcal{D}_p(u)$ it implies

$$0 \leq \int_\Omega |\nabla_{\mathbb{H}} u_\delta|^p dx - \int_\Omega |\nabla_{\mathbb{H}} u|^p dx \leq \int_\Omega \left(\delta^2 + |\nabla_{\mathbb{H}} u|^2 \right)^{\frac{p}{2}} dx - \int_\Omega |\nabla_{\mathbb{H}} u|^p dx. \tag{5.79}$$

By Lebesgue's dominated convergence theorem the last term in the previous inequality tends to zero, so we have

$$\|\nabla_{\mathbb{H}} u_\delta\|_{L^p(\Omega)} \longrightarrow \|\nabla_{\mathbb{H}} u\|_{L^p(\Omega)}. \tag{5.80}$$

This means that u_δ is a minimizing sequence for the functional $\mathcal{D}_p$, so repeating the same argument used in the proof of Theorem 3.3 we obtain, possibly passing to a subsequence, that

$$u_\delta \xrightarrow{w} u \quad \text{in } HW^{1,p}(\Omega). \tag{5.81}$$

Putting together (5.80) and (5.81) from well known results in Sobolev spaces we get $u_\delta \longrightarrow u$ in $HW^{1,p}(\Omega)$. Now we can use Theorem 5.2 to obtain estimates on $\|\nabla_{\mathbb{H}} u_\delta\|_{L^\infty(B_r)}$ independent of the non degeneracy parameter δ, namely

$$\|\nabla_{\mathbb{H}} u_\delta\|_{L^\infty(B_r)} \le C_p \left(\fint_{B_{2r}} \left(\delta^2 + |\nabla_{\mathbb{H}} u_\delta|^2 \right)^{\frac{p}{2}} dx \right)^{\frac{1}{p}}. \tag{5.82}$$

Since we have L^p convergence of the gradients we can pass to a subsequence that converges pointwisely almost everywhere and using again (5.78) obtain the desired result. □

Reference

1. Zhong, X.: Regularity for variational problems in the Heisenberg group. Preprint (2009)

Index

© The Author(s) 2015
D. Ricciotti, *p-Laplace Equation in the Heisenberg Group*,
SpringerBriefs in Mathematics, DOI 10.1007/978-3-319-23790-9